ADDRESSING EARTH'S CHALLENGES

APPLYING GIS

ADDRESSING EARTH'S CHALLENGES

GIS FOR EARTH SCIENCES

Edited by

Lorraine Tighe, PhD
Matt Artz

Esri Press
REDLANDS | CALIFORNIA

Esri Press, 380 New York Street, Redlands, California 92373-8100

Printed in the United States of America.

Second printing 2024

ISBN: 9781589487529
Library of Congress Control Number: 2023940017

For purchasing and distribution options (both domestic and international), please visit esripress.esri.com.

On the cover: Image by Ungrim.

CONTENTS

INTRODUCTION

EARTH IS FACING SOME OF THE BIGGEST ENVIRONMENTAL challenges in modern history. The changing climate presents more extreme weather hazard scenarios, including rising seas, melting sea ice, and hotter temperatures. The human footprint has reached every corner of the world, altering the climate, oceans, lands, and biodiversity in unprecedented ways. The scientific community is working to understand our role in climate change and address this existential threat by using geographic information system (GIS) technology. GIS and location intelligence are helping scientists understand climate trends and predict and assess their future impacts on the planet. The use of GIS can help decision-makers as they manage resources and create sustainable solutions. Earth science professionals use GIS to access and deliver analysis-ready data, turn raw data into useful information, and collaborate on scientific innovation to support science-based decisions and policies.

A holistic approach

Earth science professionals have found that GIS helps them address climate change in interconnected and inclusive ways. GIS offers many advantages in the context of a geographic approach:

- **Innovate with integrated geospatial infrastructure that connects data, processes, and people with GIS technology.**

- Access curated location-based, analysis-ready data in one intelligent system.
- Deliver real-time data through maps, apps, and dashboards.
- Infuse location intelligence in scientific methods using out-of-the-box modeling, environments and open-source algorithms, libraries, and coding languages.
- Facilitate research and collaborate to work across boundaries and eliminate silos.

Incorporate and access earth observations

GIS enables better science with analytics and informative visualizations. It links GIS datasets and facilitates interdepartmental information sharing and communication. With a shared database, data can be collected once and used repeatedly as one department benefits from the work of another. The use of GIS reduces redundancy, increases productivity, provides informative visualizations, and supports efficient operations. With GIS, science professionals can spend more time on analysis, discovery, and innovation.

How you can use GIS in your organization

GIS provides information to more people more efficiently than ever before through cloud-based web tools and data access services.

- **Access big data:** ArcGIS® integrates spatial analysis, statistics, demography, and other multidisciplinary data with a location as the connective thread for using curated big data. With GIS, users can manage and aggregate disparate information in a unified infrastructure for secure and timely data exchange with stakeholders.

- **Remain agile in problem-solving abilities:** GIS is open and interoperable to integrate and work with tools and languages such as Python, Jupyter Notebook, R, and out-of-the-box tools to start immediately.
- **Tackle multidisciplinary problems:** Spatial and spatiotemporal analytics and seamless integration capabilities enable multidisciplinary problem solving.
- **Build operational efficiency:** Spatial analysis and data science on demand and at scale lets you integrate out-of-the-box modeling environments with open-source spatial algorithms, data libraries, and programming languages to quickly uncover hidden patterns and transform weather services. Weather and climate data update on the scale of minutes to years, and users can combine Python with GIS to analyze the data and publish the results for others to consume.

Next-generation GIS increases our understanding of changing conditions on Earth, with better climate forecasting, monitoring, and reporting and new and broader tools, workflows, and data access.

From science to action

GIS presents a way to comprehend the world and address and communicate the challenges we face using the common language of mapping. Esri® refers to this idea as The Science of Where®.

The Science of Where helps us understand location—where things are—and apply geospatial thinking and spatial analysis for more effective action. Combining mapping and analytics in a common visual language, The Science of Where empowers the GIS community to practice open science, connecting data to agencies, organizations, and people for a larger impact for all.

From large-scale imagery analysis to collecting data in the field, ingesting big data, and analyzing real-time information, The Science of Where allows users to see and work with data in new and innovative ways, as illustrated by the stories in this book.

Stories and strategies

Addressing Earth's Challenges: GIS for Earth Sciences presents a collection of real-life stories that illustrate how earth science organizations use GIS to support geoscience, sustainable energy, environmental monitoring, climate science, weather, and marine science. The stories and strategies help readers understand how to use GIS and integrate spatial reasoning into natural resource planning and operations. The book concludes with a section about next steps with GIS, which provides ideas, strategies, tools, and suggested actions that organizations can take to build location intelligence into decision-making and operational workflows.

This book presents location intelligence as another layer of knowledge that managers and practitioners can add to their existing experience and expertise, offering a geographic approach they can incorporate into daily operations and planning. If location intelligence isn't currently part of an organization's decision-making and operational processes or used to improve constituent satisfaction, managers can use this book to start developing ideas, approaches, and skills in those areas. Developing these skills does not require GIS expertise, nor does it require managers to disregard their previous experience and knowledge. A geographic approach merely adds another perspective to think about solving problems in a real-world context.

HOW TO USE THIS BOOK

THIS BOOK IS DESIGNED AS A GUIDE TO HELP YOU TAKE THE first step with GIS to address issues that are important to you right now. It aims to help you apply a geographic approach to make decisions, improve operational processes, solve common problems, and create a more collaborative environment in your organization and community. For example, you can use this book to identify where maps, spatial analysis, and GIS apps might be helpful in your work and then, as a next step, learn more about how to apply these resources.

Learn more about the use of GIS in earth sciences by visiting the web page for this book:

go.esri.com/gfes-resources

PART 1

GEOSCIENCE

GOVERNMENT GEOLOGIC SURVEYS PROVIDE THE foundational geoscience for informing decision-support systems that allow for more significant insights and visualizations of earth dynamics. Earth observations and geoscience, coupled with location-based information provided layer by layer, can become a digital twin of the earth. GIS connects these layers within a geospatial framework to uncover value from geologic mapping. Valuable geoscience insights support sustainable resource development, land-use planning, and risk mitigation. With GIS, earth science organizations can transform disparate field activities for collecting, managing, analyzing, and sharing information. Grounded in location, geoscience GIS improves data interpretation and understanding of earth dynamics that can be shared through lightweight engagement apps.

Modern geologic mapping

A web-based enterprise GIS helps users integrate methods and data. It promotes collaboration among geoscientists and provides access to centralized geospatial data to coordinate field data collections and adopt more precise geologic mapping.

Next-generation geologic modeling

A geospatial infrastructure facilitates geologic modeling that interactively portrays complex geologic conditions at depth. Seamless

2D and 3D analyses and visualizations help users understand natural hazards, hydrology, and resource recovery.

Enterprise content access

An integrated centralized data repository underpinned by location intelligence enables everyone to find, share, and analyze the information as needed. Users can work more efficiently with consistent and accurate data exchange in secure cloud or hybrid data services.

Taking advantage of enterprise content, users can discover more significant insights into the geologic landscape. Advances in geoscience 3D visualization, prediction, and the use of apps to support decision-making offer these and other advantages:

- **Use science-based technology:** Provide an integrated, open, and interoperable environment for geologists and teams to create, access, view, and analyze geologic information using a 3D web-enabled GIS.

- **Streamline field data collection:** Transform disparate field activities with an all-in-one field app, unifying geoscience maps with digital tools for streamlined data collection, editing, and sharing.

- **Collect field data and update maps faster:** Improve geologic map compilation efficiency within distributed mapping projects using location intelligence to better collect, access, and update map data.

- **See through geologic time:** Modernize cross-sectioning and geologic interpretation of geologic features by slicing through Earth's substrata using 3D visualization and modeling.

- **Operate securely with other scientists:** Access geologic layers across systems, and quickly share results and research with colleagues using a built-in collaborative framework.
- **Deepen knowledge and abilities:** Enhance skills with ArcGIS training in spatial data visualization, 3D geologic mapping, modeling, and time series analysis to understand Earth's landscape.

GIS in action

Next, we will look at some real-life stories of how organizations are using GIS to uncover value from geologic mapping and modeling.

MINERAL EXPLORATION FROM SPACE

Exploration Mapping Group Inc.

FUTURE ADVANCES IN HYPERSPECTRAL IMAGERY PROMISE to be a boon for mineral exploration. Although remote sensing technology is improving rapidly, many satellites are not equipped to capture the quality of imagery needed to accurately decide where to look for deposits of copper ore, zinc, or other minerals.

"Very few of the satellites currently orbiting the earth can measure rock, mineral, soil, and vegetation features at the scale of interest required by the mining and petroleum industries," said Dan Taranik, managing director of Exploration Mapping Group Inc. in Las Vegas, Nevada.

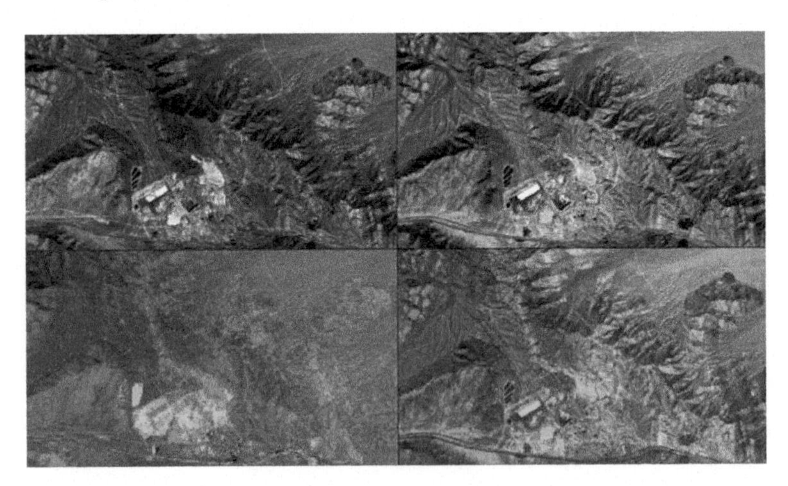

Mountain Pass Mine in California's Mojave Desert has one of the largest and highest-grade rare earth element metal deposits in the world. WorldView-3 satellite images show a natural color composite (*upper left*), an iron enhancement composite showing differences in iron mineralogy (*upper right*), a clay enhancement composite showing differences in clay minerals (*lower left*), and a lithologic composite showing differences in geologic units (*lower right*). WorldView-3 satellite image data provided courtesy of Maxar and processed by Exploration Mapping Group Inc.

Taranik has more than 30 years of experience in the mining and petroleum industries. His company specializes in providing remote sensing services to natural resource companies worldwide, often delivering finished products to customers that use ArcGIS software in formats such as map packages (ArcGIS MPKX files).

"The vast majority of satellites being launched today—known as smallsats—are simple red, green, blue, and near-infrared platforms that lack the ability to map specific clay and iron minerals that are key to discovering new mineral and hydrocarbon resources," he said. "Smallsats are fine for capturing color imagery over an area but not up to the task of detecting the specific minerals associated with copper, gold, and diamond deposits or the signs of vegetation stress in individual plants and trees."

Taranik's company primarily uses industrial-grade super-spectral instruments with 15 or more high-resolution spectral bands to capture imagery.

The mineral frontier

Remote sensing, the foundation to mineral exploration, has been used for decades to explore for minerals. Initially, aerial surveys were flown to capture images of an area where a known mineral in substantial quantities was located. These images were compared with those of other locations having similar exposed outcrops. Image analysts would examine the two sets of photographs and try to determine the likelihood that the new area would also contain the same mineral before sending expedition teams to further explore and evaluate the area.

In the post-World War II era, satellite sensor technology evolved to include radar and infrared cameras. These new sensors had advantages over conventional aerial photography because of their ability to see through cloud cover and even spot camouflage. Remote sensing analysts working with these new sources of satellite imagery, however, still relied on the same compare-and-contrast methods pioneered in the original aerial surveys: looking for areas with similar surface characteristics as known mining deposits.

Initiated by geologists from the US Geological Survey (USGS) and sent up by the National Aeronautics and Space Administration (NASA), various Landsat satellites have been continuously collecting data for nearly 50 years. In addition, about 15 other countries and agencies subsequently launched their own space missions for scientific research, with a combined total of about 5,000 satellites in orbit. Today, mining companies employ specialized companies to analyze spectral data of specific areas collected by the satellite constellations that circle the earth to help determine locations for mineral exploration and mining.

The art and science of spectral analysis

More than 4,000 natural minerals can be found on the earth, and each has its own unique chemical composition. The amount of solar radiation that a mineral reflects, transmits, and emits because of its chemical composition is like a fingerprint, or spectral signature. By measuring the tiny wavelength variations, remote sensing can identify a mineral's spectral signature from space.

"Our company analyzes the spectral imagery obtained from earth observation satellites to identify and map mineral signatures, as well as determine where oil and gas pathfinder minerals may be located for our customers," Taranik said. "The WorldView-3 satellite, for example, has such high spectral and radiometric quality that we can measure ethane and methane gas leakage in the atmosphere after the proper atmospheric corrections have been applied to the data."

WorldView-3, launched in 2014, was designed, in part, for geologic exploration. Its single panchromatic (pan) spectral band is used to rapidly collect high-resolution imagery, which is particularly

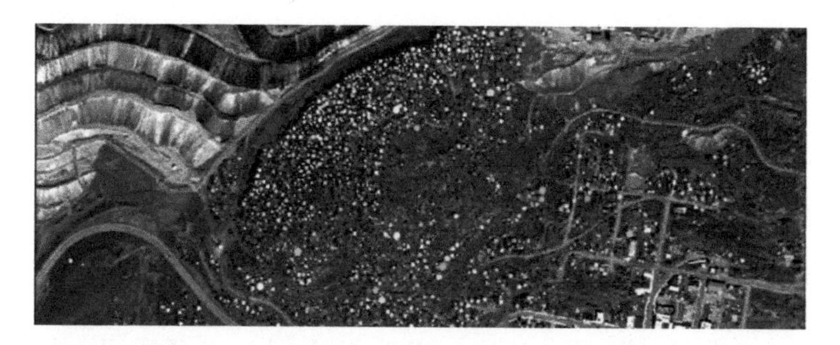

Exploration Mapping Group used WorldView-3 satellite data to measure environmental impacts of single-plant health, as shown in this community near a mine. Image courtesy of Exploration Mapping Group Inc. and DigitalGlobe.

useful for capturing sharp image detail. The visible and near-infrared (VNIR) system collects eight high-resolution multispectral bands used primarily for iron minerals, rare earth elements, vegetation health, and coastal and land-use applications. The pan and VNIR systems are complemented by eight shortwave infrared (SWIR) bands for the measurement and mapping of clay minerals and an atmospheric sensor known as CAVIS (Cloud, Aerosol, Vapor, Ice, and Snow) with 12 additional spectral bands. CAVIS bands provide accurate atmospheric corrections of the imagery for the effects of clouds, aerosols, vapor, ice, and snow.

Metallic ore deposits and their constituent minerals have characteristic properties that are visible using different wavelengths of light beyond the visible range. Those unique properties can be evaluated to map the distribution of specific minerals.

To do this, Taranik's company uses image processing approaches including spectral curve fitting; multivariate techniques; decision trees; log residuals; spectral libraries; mineral mixing/unmixing; subpixel mixture analysis; and, more recently, artificial intelligence (AI) and pattern recognition to determine which minerals are present.

Not all approaches work the same way everywhere in the world; the tools Exploration Mapping Group Inc. uses depend on the commodity of interest, geologic model, vegetation cover, and terrain characteristics.

In some especially difficult terrains for optical imagery, such as the perpetually clouded regions of Brazil, Papua New Guinea, and parts of Africa and South America, Exploration Mapping Group uses radar satellite imagery to penetrate cloud cover, Taranik said. "We are also using the vegetation cover to our advantage to measure the vegetation stress of individual plants and trees for our clients."

The original big data

Most Exploration Mapping Group customers are ArcGIS users. "We deliver our data to them in a format they can use," said Taranik. "We are longtime ArcGIS users and normally provide our finished products in multiple Esri formats—usually a map package including multiple raster formats, a geodatabase, and/or shapefiles—so the mapping layers can be easily overlain with other datasets."

Taranik said that satellite imagery is the original big data type, and his company's deliverable products are routinely over several hundred gigabytes in size. "We have found that the Esri pyramid lossless raster compression is especially useful so that large datasets display quickly on normal computing hardware, and other specialized image processing software isn't necessary to view it."

The type of satellite imagery that Exploration Mapping Group. recommends using and processes is based on how a customer plans to use that imagery.

When a customer wants to look at large regional areas, Landsat imagery would be a good choice. "If you want to map all of southern Peru, for instance, it would probably be around 30 Landsat scenes," Taranik said. "This would give you a broad view for mineral exploration—where access routes exist, [where] geological contacts [are], where rocks and soil are exposed—as well as any visible alteration in the geology that suggests where you might want to follow up with more detailed mapping, geochemical sampling, or drilling."

Exploration Mapping Group recently completed a large survey of more than 6,000 square kilometers in size for an area in the world that is, in Taranik's words, "experiencing a modern-day gold rush."

"The client wanted to see every boulder, geological contact, and concentration of alteration minerals in the vast but remote region," he said.

Taranik envisions collecting satellite data over hundreds of square kilometers of land in another area to look for unique circular features that have a kimberlitic clay mineral response in them. An igneous rock, kimberlite can sometimes contain diamonds.

Satellite imagery also can be used to detect the presence of ore in heavily vegetated terrains of Southeast Asia. "We can detect and map rock alteration and primary ore mineral signatures between the trees, in dirt road tracks, on animal and human footpaths, and the upturned soils of artisanal agricultural fields," said Taranik. "We can identify a lot of different minerals and in what quantity from space. It is opening up new regions of the world to spectral mapping and geoscientific applications that would have been missed by the previous generation of resource satellites."

The sky's the limit for satellite services

Some governments and large companies use satellite constellations such as WorldView-3 that have high data quality, with geometric precision and bit depth that provide great radiometric accuracy in their measurements, according to Taranik. "There are some governments that will be launching hyperspectral satellites in the next few years that will provide even greater hyperspectral ranges," he said.

Taranik also points to the growing number of nano- and micro-satellites that are inexpensive in terms of relative cost, although a small number are expected to fail at launch.

"The data quality is not as high as the top-end satellites," he said. "But if your application requires red, green, blue, and visible near-infrared reflectance for mapping or change detection or repeat

WorldView-3 satellite data with 30-centimeter pixel resolution provides the highest commercially available level of image detail. Colorful mining ore stockpiles, heap leach pads, and tailings dams are monitored for early warning detection of leaks, seeps, or spills. Mining haul trucks can be seen for scale. Image courtesy of Exploration Mapping Group Inc. and Maxar Technologies.

coverage of an area within a single day, you'll be able to get it from these satellites."

Sensors used by unmanned aerial vehicles (UAVs) tend to be on the lower end of the spectral capability and camera quality and are restricted to line-of-sight applications over specific mine sites for stockpile volume measurements, according to Taranik. He said that aerial surveys using traditional aircraft face challenges such as difficult remote site access, pilot and instrument technician costs, sensor import and permitting regulations, mechanical failures, and weather delays that can be complex and expensive.

"We can literally jump over these limitations with 'open skies' satellite technology, where sophisticated imagery can be acquired and processed within days or weeks of an order for any location in the world," Taranik said.

This lithologic color composite image is effective in discriminating general geology and surface cover types and shows muscovite clay-rich, hydrothermally altered granite in orange-red tones, granitic gneiss in yellow, and alkalic granitoids in blue. WorldView-3 satellite image data provided courtesy of Maxar and processed by Exploration Mapping Group Inc.

The WorldView-3 satellite, for example, orbits Earth once every 97 minutes and passes over the same point on the earth at mid-latitudes once every three to five days. The Maxar collection planning team includes dedicated weather specialists for every region of the globe who look at historical cloud cover, monsoon periods, snow cover, and any existing competition for satellite time to accurately determine the time required for collecting new imagery.

Once a client confirms an order, Taranik said it's not unusual for cloud-free images to be collected the next day.

Business opportunities in the cloud

Exploration Mapping Group uses ArcGIS technology such as ArcGIS Pro to process and enhance images and conduct vector analytics. "We also use Esri tools to create the final images, spatially validate them with the Check Geometry tool to catch any errors, and deliver them in an ArcGIS map package and geodatabase format," Taranik said. "The vector analytics are especially useful with polygon overlay and intersection to generate maps with specific exploration and environmental areas of interest based on multiple layers of input."

Taranik said Esri has sophisticated and flexible IT infrastructure models, data center deployment, portal interfaces, authentication, and distribution of data to end-user clients. Esri also is leading the migration to cloud storage and data management. "Along with the advances that are being made in data processing, this will increase the potential for continued expansion of mineral and petroleum exploration companies using remote sensing."

With ever-improving satellite technology, Taranik sees a bright future for mineral exploration.

"The latest satellite technology provides exploration geologists with the ability to rapidly screen large areas in remote places for mineral and petroleum potential," Taranik said. "It helps streamline exploration efforts and costs by reducing financial commitments for mineral leases, avoiding unnecessary ground-based environmental impacts, and ensuring labor-intensive field operations are focused on the best areas. The remote sensing tools available today are an order of magnitude better than what we had just five years ago."

A version of this story by Jim Baumann originally appeared in the *Esri Australia Blog* on February 26, 2020.

GIS PAVES THE WAY FOR A CLEAN ENERGY FUTURE

Esri and State of California

CALIFORNIA'S PLAN TO HAVE ALL NEW CARS SOLD IN THE state operate emission free by 2035 could potentially hasten the country's transition to electric cars using lithium-ion batteries. Miners, manufacturers, and logistics companies stand to benefit if the plan increases development of the state's lithium reserves.

If those industry players create a Southern California hub for lithium extraction and battery production—an outcome the state is encouraging—location intelligence will be key to making the process efficient and collaborative.

From Imperial Valley to Lithium Valley

Most lithium today comes from Australia; China; and the "lithium triangle" that covers parts of Argentina, Chile, and Bolivia. California's Imperial Valley, a desert region in the southeast corner of the state, could contain even larger reserves.

According to a study by SRI International, highlighted in *Bloomberg Businessweek*, the magma-heated brine underneath the Salton Sea could annually yield eight times the amount of lithium produced globally in 2019. With the lithium-ion battery market on track to reach $129 billion by 2027—up from $37 billion in 2019—the Salton Sea's lithium could be in high demand.

Geothermal plants already dot the shore of the inland sea, pumping the brine and converting it into turbine-powering steam. Several companies have expressed an interest in extracting lithium before the brine is sent back underground, according to *Businessweek*.

California officials hope manufacturers will build battery factories—and even plants that assemble electric cars—creating an industry cluster in a region that officials have rebranded Lithium Valley.

A unified lithium supply chain

This kind of ecosystem could consolidate a key part of the electric car industry in one area, creating supply chain efficiencies for the companies involved and benefits for the local area as well as the state.

This kind of consolidation will require cooperation among diverse stakeholders, beginning with energy companies, mining interests, local officials and community leaders, and environmental groups—potentially expanding to include battery and car manufacturers.

In other industries, such cooperation has started with a smart map of an area, stored and accessible with GIS technology. Mining companies already routinely use GIS to create detailed maps of their sites, for everything from exploration to operations. Integrating satellite imagery, remotely sensed lidar data, and even point cloud maps created by drones, these companies gain a complete picture of a project. The data helps companies identify the location of deposits, strategize how to extract them, organize the daily operations of the site, and track the environmental impact of these activities.

For any cluster that might develop near the Salton Sea, a smart map could be a platform for communications among the various interests, with cloud-based permissions governing which companies and individuals can access which data. That shared ground truth promotes transparency and can drive collaboration within industry coalitions.

Easing the environmental footprint of lithium

One consistent criticism of electric cars as clean-energy solutions is that extracting and processing the key elements in lithium-ion batteries is itself a carbon-intensive process. Extracting lithium from brine involves large amounts of water, and both requires and produces chemicals that are potentially harmful to local air, water, and soil.

Some projects in South America's lithium triangle have come under fire for misuse of local water reserves and for inattention to the environmental cost of lithium mining.

Location intelligence could help the companies pursuing Salton Sea projects maximize both efficiency and sustainability, similar to the way other resource-intensive companies are using location intelligence. For instance, GIS technology could monitor the yearlong lithium refinement process. As lithium production ramps up, the same system could help lithium-ion battery makers plan factory sites and manage the logistics of distribution.

Growing Lithium Valley

If the battery factories materialize, they will integrate locally produced lithium with other battery components sourced from around the world. Factory owners can use location intelligence to monitor their global supply chain and reduce the risk of disruption, as automaker GM has done for years.

Cobalt, another key component in electric vehicle (EV) batteries, comes from several countries that have been accused of exploiting mining labor. To ensure that they do not perpetuate bad labor practices, battery makers could follow the example of food and beverage companies that track the sources of palm oil to control deforestation.

Once the lithium in the batteries leaves Imperial Valley, a Salton Sea lithium industry could continue to exert a positive influence. Companies around the world are beginning to see their products as part of a circular economy rather than one that extends in a straight line from the manufacturer to the consumer to the landfill. When the Salton Sea lithium leaves the area in battery form to power electric cars around the world, manufacturers could take a location-intelligent approach to recycling. A manufacturer could estimate the life-span of any given battery, contact the owner around the time the battery is likely to expire, and offer buybacks. Instead of winding up in landfills, some components of the battery could be reused and reintegrated into the industrial process, creating a profitable and sustainable circular economy.

Building a lithium valley will take time, determination, and patience. But with a holistic application of location intelligence technology, an outpost of the world's clean-energy future could bloom in California's desert.

A version of this story by Geoff Wade originally titled "California Eyes Sea of Opportunity for EVs" appeared in *WhereNext* on January 19, 2021.

IMPROVING A CENTRAL REPOSITORY FOR DATA AND REPORTS

Geoscience

FOUNDED IN 1978, GEOSCIENCE IS A CALIFORNIA-BASED groundwater consulting firm that helps water utilities, government agencies, and the private sector worldwide overcome challenges to develop and preserve groundwater resources. The staff of Geoscience—composed of field personnel and teams dedicated to groundwater production, modeling, and legal projects—offers water providers data and advice to manage their water resources and defend groundwater rights.

Geoscience has collected specialized, regional data during more than four decades of being in business, but all the company's data was stored in different formats and databases. The decentralized storage made it difficult for staff to research and access historical data and reports for clients.

David Bauer, GIS coordinator and graphics specialist at Geoscience, developed a centralized database that made this vital information available in one location. The database improved operations and access to data and reports across the organization.

The challenge

Geoscience has collected and stored an array of data, including where projects are located, the client coverage of a particular area, the groundwater basin an area encompasses, and the number of wells in an area. Each of these wells has information about the ground's lithology, water level, and makeup of the soil.

According to Bauer, performing basic searches for projects and retrieving data once took hours and required several steps because the data was stored in many places. New employees faced the challenge of finding and using decentralized information gathered over 45 years, he said.

After trying some interim solutions, Bauer wanted to develop a GIS-based tool that would enable Geoscience staff to tap into the company's vast resources, regardless of GIS knowledge or experience. Because the geographic component of data is vital to the company's work, he also wanted the new tool to link to Geoscience's PDF reports for easier access.

The solution

Bauer found a 2016 YouTube video discussing ArcGIS for SharePoint®, which allows users to map and share content to visualize, organize, and interact with data. One feature stood out to Bauer: the ability to link PDF files to a location. Users can click on a location in an interactive map display to retrieve a report. He believed ArcGIS for SharePoint could be the answer to many of the company's needs.

Bauer chose ArcGIS for SharePoint to develop a GIS-based tool to help staff research the Geoscience database. He says he set up a "map-based tool for future research and basic analysis for all users in the company, in a secure company-wide environment."

He liked many of the specific features found in ArcGIS for SharePoint:

- Tools for searching company databases in one location
- Graphic ability for displaying data easily
- Ability for geotagging, or linking, PDF files to point or polygon location data
- Capability of using the vast ArcGIS Living Atlas of the World databases

The connectivity ArcGIS for SharePoint offers was a bonus. Bauer began uploading data into the SharePoint environment so that field personnel could access reports from anywhere with an internet connection.

Bauer incorporated Geoscience's work databases and reports into ArcGIS for SharePoint. He set up the program as an easy-to-use, map-based, searchable research tool. He produced demonstration videos and instructions within the company's secured online SharePoint environment to help train staff.

Users can select a map, zoom to a specific area, and view the PDF report as a pop-up. They can also search the different shapefiles and informational databases that are associated with it. The current SharePoint site houses tens of thousands of items.

"It's always good to learn and grow, and so this was a wonderful way to kind of readjust and make [our system] more user-friendly for the office," says Bauer.

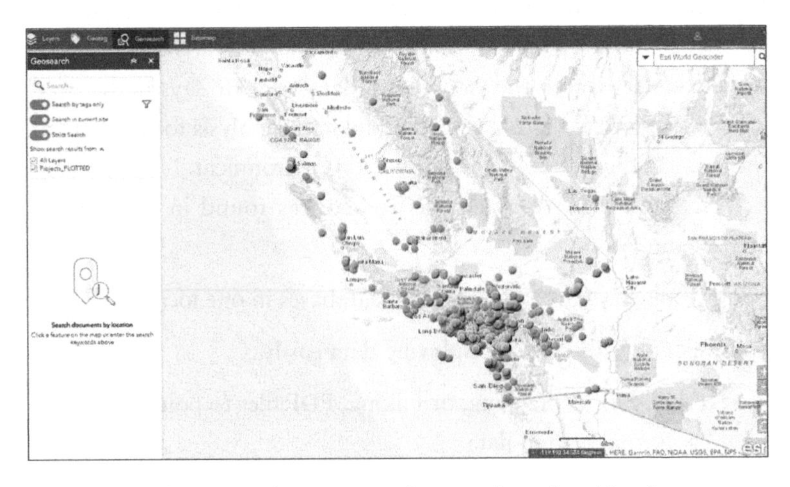

Map showing locations of assets spread across Central and Southern California.

The results

The use of ArcGIS for SharePoint improved daily operations at Geoscience, including how staff research and access data. As a result, research that once took hours now only takes a few minutes because reports and data are just clicks away, saving time and reducing overhead costs.

"In the consulting industry, time is money. This tool makes the staff more efficient when performing research or processing information in a graphical format," Bauer said. "They can fulfill the client's request literally within minutes."

Bauer says the relative simplicity ArcGIS for SharePoint is a significant benefit, because only one-third of the staff is familiar with GIS.

With a small learning curve for the user, anyone can use the software. "It opens an entirely new, map-based perspective to your stored data," he said.

The new solution has also improved Geoscience's field data collection workflows. Bauer said having everything available in SharePoint means that all data is online and accessible in the field via phone, tablet, or laptop. Also, since reports can be linked to a point or polygon, staff in the field can get the information they need from anywhere, allowing them to respond to client needs faster.

Bauer explained that using cloud-based GIS mapping software ArcGIS Online in tandem with Geoscience's SharePoint databases allows the team to quickly access internal data. It also serves as an efficient search tool for obtaining information about previous projects and experiences.

"I use [ArcGIS Online] for planning purposes since I can easily identify the number of wells owned and operated by water providers in a given region. It also helps quickly identify where our projects are located in relation to a new client or potential project," says Sean

Stewart, Geoscience marketing director. "Whereas before, it would take a few hours to research past projects and create a list or a map, we can now identify projects in minutes with just a few clicks."

Newer employees now have immediate access to decades of institutional knowledge, said Chris Coppinger, senior geohydrologist at Geoscience. Staff members no longer need to redo work that's already been done because they can quickly locate work performed in a groundwater basin, near a plume, or near a specific city, he said. And the software provides immediate templates for future work.

Bauer looks forward to future software enhancements to further improve Geoscience's research capabilities and produce the best and most efficient work for the company's clients.

A version of this story titled "Groundwater Consulting Firm Improves Its Central Repository for Data, Reports with Mapping Technology" originally appeared on esri.com.

BRINGING SCIENCE STORIES TO LIFE WITH GIS

San Diego State University

SCIENCE AND JOURNALISM HAVE A LOT IN COMMON. BOTH fields require practitioners to ask questions, explore data, analyze information, and then act on the knowledge—usually by preparing a final report, whether it is a scientific paper or a news article.

At California's San Diego State University (SDSU), students from the journalism and geology departments collaborated on a 15-week live news and science experiment to explore the air quality in four San Diego neighborhoods: Barrio Logan, Logan Heights, Chollas Creek, and Bankers Hill. Funded by a grant from the Online News Association, students built sensor kits (based on open-source technology) that measured particulate matter and different gases—such as liquefied petroleum gas, isobutane, methane, and smoke—in the air. They used ArcGIS software to map and analyze the information collected by the sensors. In the end, the students produced nine news stories and two videos for *inewsource*, an independent, data-driven online news organization based in San Diego.

The course outline was grounded in the geographic inquiry process espoused by Esri, which encourages students to ask questions, explore and analyze data, and then act on their findings. By engaging in this process, students advanced their scientific and journalistic skills. They also learned that scientists, journalists, and the public can use GIS as a tool to discover details and share stories about the environment.

What is in the air?

Bounded on several sides by mountains and beset by dry air and freeway congestion, California's fourth-largest maritime port, San Diego, tends to attract and retain pollution.

The American Lung Association's *State of the Air* report for 2015 ranked San Diego 35th out of 180 cities for the number of days with high ozone pollution, or smog. The report also ranked the San Diego area 39th for the average amount of particle pollution (a medley of extremely small liquid and solid pollutant particles) present during a 24-hour period and 40th for the levels of particle pollution in the air annually. This ranking is significantly better than in 2011, when San Diego ranked seventh for ozone pollution and 15th for short-term particle pollution.

To sustain or even accelerate improvements in air quality, it helps to know what is in the air. From early February to late April 2015, students in SDSU's digital journalism class monitored the air quality in different parts of San Diego with the ultimate objective of better informing the public about the area's pollutants and their effects.

Mapping and visualizing pollution

To begin the project, students assembled environmental sensor packages developed by SDSU staff and a sensor consultant using open-source technology. Each kit included sensors that pick up airborne particulate matter and gases. They each also had an Arduino board that read the sensor information every 30 seconds. This small computer then sent the data to an LCD display, which showed details about the presence and concentration of different pollutants in the air. The data collected by each kit was stored in a mini SD memory card in the Arduino board.

After each of their eight data collection trips (where they went to the same spots on the same day of the week at the same time), students took their mini SD cards and put the recorded data—along with the longitude and latitude where each recording was taken—into a spreadsheet. Then they imported the tabulated data into ArcGIS Online to map it.

Students quickly learned they could more easily visualize data once their spreadsheets became maps. Simple GIS tools allowed them to examine patterns in the environment and trends in various neighborhoods, and then communicate these stories with their maps.

Bringing stories to life

GIS helped students find and bring their stories come to life. Several students presented their accounts as news articles on inewsource.org.

As one student reported, in the neighborhood of Barrio Logan, which is sandwiched between the Interstate 5 freeway and the Port of San Diego, high volumes of particulates were discovered, especially from black carbon, which, according to the article, is the most noxious fine particle. With little wind to blow pollution away and the additional strain of having the San Diego Coronado Bay Bridge overhead in some areas, this neighborhood has one of the worst air quality ratings in all of San Diego.

Students also discovered that Logan Heights, just north of Barrio Logan, is afflicted by similarly low-quality air. Located between two freeways, its residents—especially children—are at elevated risk of developing and exacerbating respiratory illnesses such as asthma.

Bankers Hill, however, fared better. Students reported on inewsource.org that although the neighborhood sits directly to the east of San Diego International Airport and is also circumscribed by freeways, this hilltop district receives constant wind—enough to break up pollutants. That said, some areas of the neighborhood that are closer to the freeway experience an eddy effect, meaning that the wind recirculates polluted air.

Underlying all these observations were the students' ArcGIS Online maps, which made the data meaningful and animated the science.

The sky is the limit

The sensor journalism project engendered a collaborative environment. It brought together students from different majors, teachers and faculty from various SDSU departments, and journalists from the community. GIS served as the connecting foundation for everyone's exploration and research.

With ArcGIS Online, the class collaborated to collect primary-source data, use existing GIS databases to understand and investigate the environment, and create storytelling maps.

The air quality information displayed on the class's final map, which amalgamates all the pollution recordings the students took in their assigned neighborhoods, supplemented existing air quality data. It guided the students as they investigated further, studying wind patterns and conducting interviews. It also helped them hypothesize about and analyze pollution patterns, sources, and impacts in San Diego's neighborhoods.

The SDSU staff involved in this project presents an interdisciplinary curriculum that brings together geology and journalism students to learn how GIS can contribute to scientific investigation and news reporting. The use of cutting-edge technology, open-source tools, and an experimental approach showed students how much they could learn through collaboration in a relatively brief time.

A version of this story by Amy Schmitz Weiss originally appeared in the Summer 2016 issue of *ArcNews*.

ENSURING THAT MINING OPERATES RESPONSIBLY AND EFFICIENTLY

New Mexico Energy, Minerals, and Natural Resources Department

NEW MEXICO, AT 121,000 SQUARE MILES, IS THE FIFTH-largest state in the United States. Overseeing mine operations throughout such a large area while staying within budget is no small achievement. But the staff of the New Mexico Energy, Minerals, and Natural Resources Department's Mining and Minerals Division (MMD) is resourceful and supplements operating costs by using GIS, acquiring data at no or low cost, and forming geospatial or technical data-sharing partnerships.

The MMD endeavors to conduct responsible mining operations, from exploration to reclamation. Maps provide a baseline for analyzing activities and disturbances made by mining operations across the state's vast landscape. MMD uses Esri GIS software to process mining operation and exploration permit applications and to report economic impacts.

Two of the four MMD programs (Coal Mine Reclamation and Abandoned Mine Land Reclamation) were created as part of the 1977 Surface Mining Control Reclamation Act (SMCRA), which formed partnership arrangements with the US Department of Interior Office of Surface Mining Reclamation and Enforcement (OSMRE). Grants provide MMD staff with the means to collect geospatial information. The Coal Mine Reclamation Program developed relationships with mining operators to share maps or geospatial data. Staff from the Mining Act Reclamation Program collect permit locations and track reclamation, using GPS to populate a geospatial database.

To assess mining reclamation operations, the state uses digital elevation models (DEMs) and digital terrain models (DTMs) from various time periods, most of which were created by the USGS. In addition, staff use orthoimagery acquired from federal, state, and local governments and by mine operators. In recent years, the MMD has used orthoimagery acquisitions from the US Department of Agriculture National Agriculture Imagery Program to create newer statewide DEMs and DTMs.

DEMs and DTMs offer geographic representations of what areas were like prior to mining. The models reveal stock tanks and dams along drainage corridors and identify those that have been breached. The models also depict patterns in drainage basins that indicate sinuosity, which provide a method for analyzing hydrologic and topological characteristics. MMD uses the digital models to ask mine operators to reproduce the degree of sinuosity that has been determined.

GIS helps MMD track mining activity throughout the state as well as enforce reclamation regulations for surface mines and abandoned mine lands. GIS indicates land change, maps mine impacts, provides guidelines for mine reclamation projects, and tracks all mandated environmental and cultural assessments before project design.

The New Mexico Environment Department also has a stake in making sure mines operate conscientiously. GIS supports MMD's collaboration with the Environment Department in considering permit applications. Maps make it easier for the agencies to review and comment on mine permits and closeout plans and ensure that environmental standards are included in each application. The Environment Department also works with MMD to monitor mining reclamation activities.

"GIS helps the Mining and Minerals Division prioritize where it should spend money on surface reclamation projects," said Linda S. DeLay of MMD.

DeLay used GIS for prioritizing cleanup activities. She analyzed the location of legacy uranium mines and ranked their priority for reclamation. The basis for assigning rank to each of these mines was its proximity to streams, agricultural sites, urban areas, and wells. DeLay then used a weighted overlay GIS model to map reclamation priorities. She presented the New Mexico Legacy Uranium Mines map at MMD and sister agency meetings and at a national conference. The map helped decision-makers decide where to allocate resources.

To monitor coal mine reclamation, MMD applied for a grant from the OSMRE Western Region to acquire WorldView-2 satellite imagery. Using this imagery along with on-the-ground vegetation surveys, the remote sensing analyst is creating vegetation change detection maps for the Vermejo Park Ranch abandoned coal mine town reclamation project to aid specialists in evaluating revegetation and wetland mitigation.

Geomorphic reclamation activities included redistributing and burying the coal waste and reforming stream channels to a more natural pattern. The purpose of this geomorphic work is to eliminate movement of waste into watershed drainages. GPS devices were attached to earth-moving equipment to map the new terrain design. The imagery was so detailed that an analyst could actually tease out how many pinion and juniper trees were in the area at the time.

Analysts also use lidar data. While satellite imagery provides a close-up picture of surface mines, GIS renderings of lidar data offer a highly detailed 3D perspective. These sophisticated 2D and 3D maps reveal the condition of an area prior to the commencement of mine operations.

MMD also used GIS to assess the impact on the vegetation and terrain around the El Segundo coal mining operation and proposed mine. To document baseline landform conditions, New Mexico Energy, Minerals, and Natural Resources Department acquisitioned

A camera mounted on a UAV rapidly takes pictures. These images go through photogrammetric processing to generate a point cloud for a 3D representation of a stream.

two 25-square-mile areas of lidar data that had been captured prior to mining operations. Staff used first-return lidar data to render vegetation density images and bare-earth lidar data to model the terrain. The lidar rendering is a blueprint for what the mining company will need to do to restore the terrain's original contour and reestablish vegetation to its initial condition.

Staff found a less expensive way to capture data by using a UAV. A compact camera mounted to a Trimble UX5 fixed-wing UAV takes overlapping orthophotos. Photogrammetric processing of these images generates a point cloud of x,y,z values rendering a 3D topographic model. MMD used this technique and GIS to create a topographic model of a stream restoration project at a historic coal mining town. By attributing the points with the photo's RGB values, staff calculated the heights of the vegetation. To spot-check the accuracy of the remote sensing data, staff went into the field to measure the heights of a sample of vegetation for comparison.

MMD currently uses Microsoft SQL Server, integrated with ArcGIS Server, to manage most of the mine information, geodatabase, and web map applications. The division is transitioning more of its geospatial data to ArcGIS Server. The geodatabase includes data from the state's resource GIS clearinghouse, mine operators, and elsewhere as well as data generated in-house.

MMD makes mining information available to the public through its website. The site is GIS enabled by ArcGIS Server with applications developed in Silverlight. Users can see the locations and names of mines, which are coded as active mines, inactive mines, and mines where bond has been released. They can also see coal mine permit boundaries, coal districts in New Mexico, and US coalfields characterized by coal type. The map also has a soil type layer and geologic period layers.

A version of this story by Barbara Leigh Shields originally titled "GIS Ensures that Mining Operates Responsibly and Efficiently" appeared on esri.com.

PART 2

SUSTAINABLE ENERGY

AS CONCERNS ABOUT CLIMATE CHANGE GROW, ENERGY organizations remain committed to creating sustainable energy sources and reducing carbon emissions. GIS technologies are helping energy leaders rebalance their portfolios to include renewable sources, net-zero emissions strategies, and climate change risks and opportunities in business decisions. The development of renewable energy sources—such as wind, solar, hydrogen, and geothermal energy facilities—and the infrastructure to support them are inherently spatial in nature. Trends in renewable energy projects are using geospatial analysis to help optimize energy transmission systems, changing how companies use renewable resources.

Achieve substantial energy practices

GIS analysis provides location intelligence that helps energy producers

- make sense of vast amounts of solar, meteorological, socioeconomic, and energy grid data to address renewable energy challenges,
- present an accurate picture of current conditions and allow users to model future strategies because it can present many data layers on a digital map, and

- enable new energy production by identifying sites with the most potential energy and economic development while minimizing environmental impacts.

Improve workflows and collaboration

Energy producers find that the capabilities of GIS facilitate research, planning, and economic development for the energy grid of the future:

- **Get science-based insight:** Deploy reliable technology that supports energy efficiency, greater energy security, and resilient and green buildings for a better quality of life.
- **Access big data quickly:** An integrated centralized data repository makes it easier for everyone to find, share, and analyze information.
- **Bring clarity to troves of data:** ArcGIS, vast data, and 2D and 3D tools enable deeper analysis of environmental, commercial, physical, and social constraints of potential energy sites.
- **Streamline mobile field collection:** ArcGIS optimizes field data collection with improved coordination, moves from paper to digital field entries, eliminates redundancies, and increases efficiencies.
- **Work together better:** Increases team and stakeholder engagement, communication, and data sharing.
- **Create collaborative plans:** Uses a framework for iterative energy planning scenarios informed by predictive modeling and endorsed through stakeholder engagement to support informed outcomes.

GIS in action

Next, we'll look at some real-life stories of how organizations are using GIS to unlock more significant renewable energy potential and support innovative clean energy strategies.

WEB MAP BRINGS TOGETHER CONSERVATION AND GREEN ENERGY DEVELOPMENT

The Nature Conservancy

THE MIDWEST IS KNOWN AS THE WIND BELT OF THE UNITED States for good reason: nearly 80 percent of the country's current and planned wind energy capacity exists in the Great Plains, an area that extends east of the Rocky Mountains and runs from northern Montana to southern Texas. The US Department of Energy's Wind Energy Technologies Office (WETO) reports that wind energy reduces CO_2 greenhouse gas emissions while also providing health and climate benefits through reduced pollution.

Much wind energy development is occurring—and is expected to increase—in the wind belt. But as wind energy developers plan new sites, they all face this question: How can companies site new wind turbines in places that are optimal for wind resources and transmission without harming wildlife or encountering costly delays from regulatory or legal challenges?

Wind projects sited in the wrong place can threaten some of the best wildlife habitat. The Nature Conservancy (TNC) estimates that renewable energy development could adversely affect as much as 76 million acres of land in the United States—an area about the size of Arizona.

An Esri technology-based resource developed by TNC can help focus renewable energy in the right places—windy areas that pose a relatively minimal risk to wildlife and their habitats. Originally called *Site Wind Right* (later expanded to include other types of renewables and renamed *Site Renewables Right*), this interactive online map is

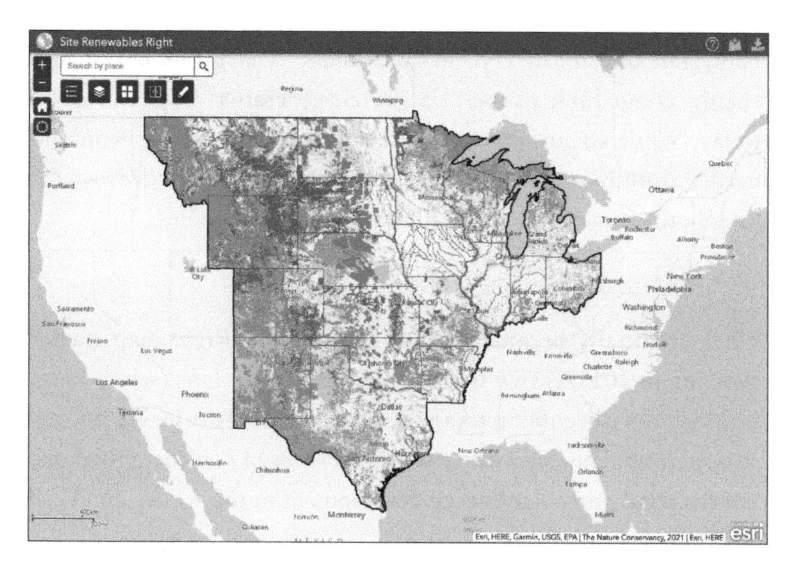

Site Renewables Right (originally called *Site Wind Right*), from The Nature Conservancy, evaluates more than 100 datasets from 17 states and shows that 76 million acres in the US wind belt could be developed for wind energy without affecting key wildlife habitats. The left part of the map is color coded to reflect key wildlife areas; the bright-green overlay on the left and right represents areas of low-impact wind development potential. Map courtesy of TNC.

available for wind developers, power purchasers, utilities, companies, state agencies, and municipalities to help reduce conflict between wind energy and conservation.

TNC developed the *Site Renewables Right* map for 17 states in the Midwest, pulling from more than 100 datasets on wildlife habitat and land use to help highlight areas with the lowest potential for environmental friction. The results of this analysis, performed by TNC scientists, are enlightening and encouraging.

"We were thrilled to discover we could generate more than 1,000 gigawatts of wind power in the central [United States] solely from new projects sited away from important wildlife areas," said Mike

Fuhr, state director of TNC in Oklahoma. "That's a lot of potential energy, comparable to total US electric generation from all sources today. While advancements in transmission and storage would be needed to fully realize this wind energy potential, it proves we can have both clean power and the land and wildlife we love."

Great potential for wind in the Great Plains

What eventually became the *Site Renewables Right* map started evolving in 2011 as two things were happening. First, wind energy facilities were beginning to appear across the Great Plains. Second, studies from TNC and other scientists showed considerable potential for wind and solar energy development in the western and central United States.

The Great Plains is home to the country's largest temperate grasslands, which are among the most altered and least protected habitats in the world. The last expanse of this once-extensive ecosystem is found in the greater Flint Hills region of Kansas and Oklahoma. Poorly sited wind turbines in places such as Flint Hills threaten wildlife such as bison, bald eagles, and the once-common greater prairie chicken that depend on this endangered and beautiful place. Turbines have the potential to harm wild animals "directly, via collisions, as well as indirectly due to noise pollution, habitat loss, and reduced survival or reproduction," according to the USGS.

But as studies demonstrated, the Great Plains could provide clean, renewable electricity without compromising wildlife habitat and other natural resources.

"Those studies showed very positive results that we can meet or exceed renewable energy goals by using sites that were previously disturbed or had relatively low conservation value," said Chris Hise, associate director of conservation for TNC in Oklahoma.

TNC scientists created a resource that energy planners could use

The temperate grasslands of the Great Plains—home to bison and other wildlife—are among the world's most altered and least-protected habitats. Photo courtesy of Chris Helzer, TNC.

early in the siting process to avoid impacting wildlife and delaying their projects. TNC is among many organizations that want properly sited wind, solar, and other renewable energy projects to succeed to meet the challenges posed by climate change.

With support from partner organizations, TNC gathered data on wildlife, habitats, land-use restrictions, and areas of biodiversity significance and organized it in ArcGIS Desktop using ArcCatalog™. With ArcMap™ and ModelBuilder™, TNC then assembled multiple spatial data layers of wildlife habitats and potential engineering and land-use constraints. Finally, using ArcGIS Web AppBuilder, the team created an online resource to share the data in what became the *Site Renewables Right* interactive map.

One of the biggest surprises for Hise and his team was the impressive number of low-impact areas across the central United States that the analysis identified—approximately 90 million acres.

Planners in the early stages of establishing a wind energy operation can study site-specific details, explore the *Site Renewables Right map*, consult with appropriate state wildlife agencies, and use the Wind Energy Guidelines developed by the US Fish and Wildlife Service to find spots that work best for everyone.

Although transmission and storage technology need to improve, the low-impact sites in the Midwest are well distributed.

"If we plan carefully, there's plenty of space to go big on wind energy in this part of the country," said Hise.

Broadening the reach of wildlife-minded green energy projects

The online map has the potential to reduce the risks of wind deployment delays and cost overruns by helping developers locate sites that are less likely to face regulatory or legal challenges. This capability has spurred the endorsement of Evergy, an energy provider in Kansas and Missouri that became an early user of the analysis.

"*Site Renewables Right* is an invaluable resource that helps us avoid unnecessary impacts to the wildlife and iconic landscapes of the Great Plains while also allowing us to provide clean, low-carbon energy for our customers," said former Evergy CEO Terry Bassham.

The mapping analysis has also invited accolades from another early reviewer, the Association of Fish & Wildlife Agencies. Additionally, the web map has received endorsements from several conservation groups, including the National Wildlife Federation and the Natural Resources Defense Council.

"We need more resources like this to speed up our move away from burning fossil fuels," said Katie Umekubo, a senior attorney at the National Resources Defense Council. "Well-sited wind energy allows us to meet our climate goals, advances conservation, and ensures that we avoid irreversible environmental impacts."

The once-vast greater prairie chicken population has fared poorly as its grassland habitats have been converted to other uses. Photo courtesy of Harvey Payne, TNC.

Currently, TNC is looking to broaden the reach of the web map within communities, companies, and government agencies so they can all apply this wildlife-minded strategy quickly and get the blades turning on clean and homegrown energy in the Great Plains.

"The Nature Conservancy supports the rapid acceleration of renewable energy development in the United States to help reduce carbon pollution," said Fuhr. "We are looking forward to providing *Site Renewables Right* to the people making important decisions about our nation's clean energy future."

A version of this story by Eric Aldrich originally titled "Web Map Brings Together Wildlife Conservation and Green Energy Development" appeared in the Fall 2020 issue of *ArcNews*.

AN INDUSTRY ON THE VERGE: GREEN HYDROGEN

Esri

AS NET-ZERO PLEDGES MAKE THEIR WAY TO THE TOP OF corporate priority lists, industry leaders are setting their sights on clean-burning hydrogen as a new ally in the decarbonization quest. Today, only a small fraction of global energy use consists of hydrogen, but recent announcements suggest growth. The location technology already used in adjacent industries could accelerate the efficiency needed for hydrogen's ramp-up.

The US Department of Energy announced plans to expand the hydrogen industry within the next decade and make it less dependent on fossil fuels, coinciding with initiatives from a European energy coalition, the Chilean government, and Japanese engineers.

Hydrogen—typically converted to energy through fuel-cell technology—could meet up to 24 percent of the world's energy needs by 2050, according to a 2020 report from BloombergNEF. But energy producers must do more. Green hydrogen produced with renewable energy currently accounts for just 0.1 percent of global supply. And it's expensive, costing up to US$6 per kilogram versus $1–3 per kilogram for conventional hydrogen. To make green hydrogen production more cost-efficient, the energy industry has begun using technologies such as virtual 3D models, analytics tools, and real-time dashboards—all forms of location intelligence—to reduce cost and increase production.

Which color is your hydrogen?

Not all hydrogen is created equal. While all hydrogen production can benefit from location intelligence, net-zero advocates favor green hydrogen, among several varieties that are given color codes, although visibly they appear colorless as an invisible gas:

- **Gray hydrogen:** This production method is by far the most common today, deriving hydrogen from the burning of natural gas. This creates nine parts of carbon dioxide for every one part of hydrogen.
- **Blue hydrogen:** Companies producing blue hydrogen also burn natural gas, but use carbon capture technology to mitigate carbon dioxide emissions.
- **Green hydrogen:** The most sustainable method of producing hydrogen, this process uses energy from wind, solar, or another renewable source to split water into hydrogen and oxygen, producing no carbon emissions.

The hydrogen resulting from these processes is clean-burning and often used in fuel cells to produce energy.

For hydrogen producers, the cost of going green is determined by one essential factor: a hydrogen plant's location. Green hydrogen can be expensive to produce through renewable energy if input costs are high, and it can be expensive to store and transport. Energy companies entering the green hydrogen market must first understand location-based variables such as regional access to renewables, local demand for hydrogen, and nearby infrastructure for hydrogen transport (for example, shipping versus pipelines).

These variables inform a series of decisions—whether to convert existing plants or construct new ones, which type of renewable energy should power the electrolysis process that produces hydrogen, and whether to store and distribute the hydrogen as a gas or a liquid.

As leaders in green hydrogen consider where and how to lower the cost of the gas, they can learn from energy companies that have relied on location technology for years. BP, for instance, uses GIS for everything from pipeline management to market research. In renewables, Renewable Energy Systems and Norway's Equinor use GIS for site selection and other processes. Henrik Hagness, leading engineer for mapping at Equinor, says spatial intelligence is "needed throughout [the process], from business development and siting to moving ahead with a development project to efficiently maintaining and inspecting the energy assets when in operation."

In growing industries such as green hydrogen, GIS-based smart maps and dashboards can provide single-pane-of-glass visualizations that identify cost-effective options and evidence-based next steps.

Bringing data together drives smarter improvements

Once green hydrogen systems are in place, energy companies will need to fine-tune operations and apply efficient processes at scale. Reliable growth is best achieved by equipping executives with a decision-support system—an operating picture based on analytics that reveals patterns and outliers and models potential outcomes.

For green hydrogen producers, key performance indicators may include a plant's renewable energy consumption rates, local weather patterns, buyer demand, and equipment conditions. For executive decisions, these datasets are best viewed together instead of apart. For example, comparing weather patterns and buyer demand—an analysis that natural gas operator ONEOK performs with GIS—helps the company determine the best time to run electrolysis.

Location analytics tools combine these pieces of information and identify connections. Visualizing the data on map-based dashboards shows where even small operational adjustments will have the most impact. Green hydrogen companies may find inspiration in energy firms that have used decision-support technologies such as GIS to integrate analytics and visualization and ensure that executives can access complete, current information.

Looking ahead with real-time awareness

As the green hydrogen industry expands, the demand for day-to-day operational efficiency will increase. Tracking equipment maintenance, clean energy certifications, and safety protocols will require a location-based view of assets and activities. For hydrogen producers, sensors and real-time data feeds will supply accurate and near-real-time information on those conditions and help avoid costly downtime.

Combining that data with maps and dashboards for visualization, companies can even create a digital twin—a virtual model of a facility or supply chain that links information to location—showing executives what's happening and where at any moment.

The energy sector already considers technology that delivers intuitive access to operational information as an essential business system. Leaders can turn to maps, dashboards, and digital twins populated by real-time data to manage the production of clean energy and more clearly focus on a sustainable future.

A version of this story by Alessandra Millican and Geoff Wade originally appeared in *WhereNext* on August 17, 2021.

LOCATION INSIGHTS POWER THE SOLAR AND WIND ENERGY INDUSTRY

Aegean Energy Group

WIND AND SOLAR ENERGY DEVELOPERS LOOK TO AEGEAN Energy Group to improve alternative energy construction and operations. Its Maps to Megawatts solution supports development, on-site analysis, site control, and reporting. The company must collect data to deliver accurate information throughout the life cycles of its customers' energy projects. Maps to Megawatts is a subscription-based online mapping and analytics technology that uses ArcGIS to deliver a scalable and secure solution from an energy project's conception to decommission. Aegean Energy extended the capability of Maps to Megawatts by enhancing it with GIS tools that streamline the collection of field data and improve location analytics.

The challenge

Maps to Megawatts helps clients with planning, strategy, and development. Aegean Energy offers suitability analysis for developers looking for wind and solar energy opportunities and identifies any costly or challenging development obstacles, such as safety, civil, or environmental issues. Aegean Energy needed a more efficient way to gather data for analysis and promptly make results available to stakeholders.

"Field information is important to us internally," says Woody Duncan, senior vice president of Aegean Energy Group. "It also helps clients be strategic and methodical in project development, which is a time-saver and critical in this day and age. We have to have greater insight as we look for energy opportunities."

The process for receiving information from the field for analysis needed retooling. Aegean Energy gathers data about a project during construction, but processing the data was slow. From project photos that required cataloging and filing to paper forms that staff manually added to a Microsoft Excel database, the workflow delayed the distribution of information from the field. The company estimated that these burdensome and slow processes took around three months to complete.

Aegean Energy also sought a more efficient way to conduct analysis and quality assurance/quality control (QA/QC) when analyzing data.

The solution

The company implemented ArcGIS Insights℠ to streamline data input, analysis, and reporting. This analytic software allows users to perform data analysis, document their workflow, and share analysis results with others. Aegean Energy uses Insights throughout the stages of the assessment process, including quality control, tracking schedules, budgets, monitoring ongoing problems in the field, and data reporting.

Maps to Megawatts functionality includes construction monitoring that tracks inspection status, which is based on data on forms completed by field operations teams. These forms help Aegean Energy identify potential development issues. Once an operator submits a form, GIS processes it. Insights flags tasks that are open and ready for inspection. Inspectors have a visual representation that shows where work must be done. Using an app to update the task status, inspectors can instantly close the loop, and Insights tracks the change. This simple workflow has improved inspection efficiency.

Aegean Energy also uses the Insights Link Analysis tool to trace form entry errors, another aspect of QA/QC. Using a network of interconnected links, the tool identifies and analyzes relationships

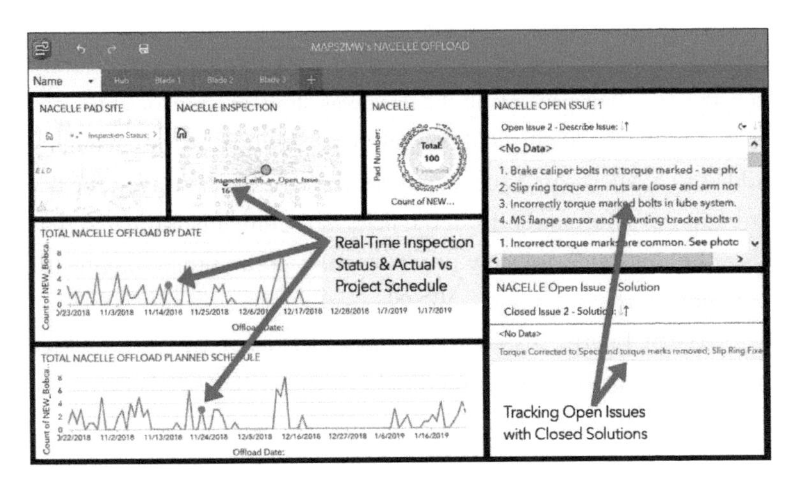

Aegean Energy uses Insights to compare real-time asset inspections with its projected schedule. Additionally, open issues are returned when users click a particular asset. This visual analysis reduces manual work.

not easily seen in raw data. Suppose that a fieldworker inaccurately enters two locations on an inspection form. Since each inspection status task links to a specific solar pad number or a wind turbine location, only one form can be entered for each asset. A link chart, with a clear graphic representation, shows a pad number with two statuses, which indicates data entry error.

"Using this tool saves hours of work," Duncan said. "Previously, somebody would have to spend many hours trying to dig through the reports and figure out if information had been entered twice or if there was another issue. But with Insights, we can immediately find the problem. It's like pulling a needle out of a haystack in an instant rather than months later."

Aegean Energy also uses other ArcGIS products to publish analysis results and expedite reporting. After staff conduct a suitability analysis, they make the results available to web clients or accessible through an ArcGIS Online web service. Companies use the results for constructability assessments in the field.

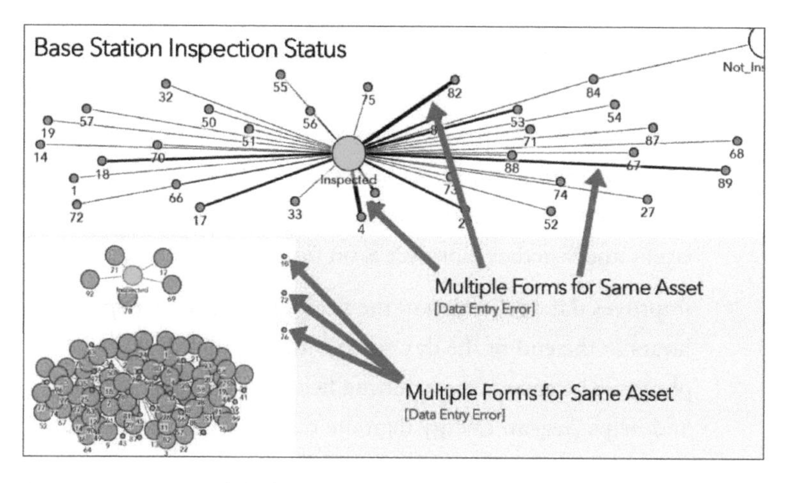

Aegean Energy uses link charts within Insights to quickly identify data entry errors. The presence of a thick connected line or an unreported inspection status signifies a data error.

Aegean Energy has further increased data capture efficiency using Esri apps—ArcGIS Collector and ArcGIS Survey123. Collector is designed for accurate data collection, while Survey123 facilitates digital survey distribution for simplified data collection. Staff use these tools to collect and organize data, connect it to the office, and share it with stakeholders.

"To have the ability to go through an enormous amount of information and simplify it into what we need to see is terrific," Duncan said.

Insights supports the company's work in these and other important ways:

- Ensures accuracy of project-related documents. On-site workers and administrative staff log in to the application and view completed forms to ensure correctness.

- Supports collaboration by allowing stakeholders—no matter their location—to view data and project-related developments. With web-based data and results available in one location, project managers, executives, site administrators, and other stakeholders can see development issues and whether a project is on time and on budget.
- Improves data collection in the field. Staff no longer spend hours at the end of the day sorting and sending forms and photos. The ease of transmitting field data improves efficiency and helps Aegean Energy digitally transform its processes.

The streamlined workflows and location-intelligence tools benefit staff. Fieldworkers use mobile devices to complete forms and share information and results on the web. Viewing project issues in Survey123 and tracking task status and resolution have brought value to customers and the company.

By enhancing the Maps to Megawatts solution with tools from Esri, Aegean Energy Group offers energy companies greater efficiency and understanding for conducting operations and developing construction for wind and solar power.

"For us, we live in maps," Duncan said. "Everything we do is tied to where, so the spatial data we gain from Insights is incredibly valuable; otherwise, it's just data."

A version of this story originally appeared on esri.com in 2019.

MAPPING RENEWABLE ENERGY POTENTIAL

Prediction of Worldwide Energy Resources team, NASA

NASA HAS FLOWN ASTRONAUTS TO THE MOON AND ROVERS to Mars, among its many space explorations. But NASA also observes Earth from space to understand our planet's interconnected systems.

NASA scientists study severe weather, natural hazards, and global food production using satellite observations and modeling. They use GIS to deliver data to users. They also estimate and monitor solar radiation and winds—data that is in high demand as the world transitions to renewable energy.

Through its Prediction of Worldwide Energy Resources (POWER) project, NASA provides interactive maps, applications, and data services that describe meteorological conditions and solar radiation and show how those resources may be changing over time. The majority of data products update on a daily basis, and records date back over 35 years. Decision-makers across industries can access the data inside their own applications to visualize on a map, perform analytics, and employ calculations to enhance planning and problem-solving.

For instance, leaders of cities and utility companies rely on NASA data to calculate solar and wind power potential for prospective power generation sites. Facility managers use it to examine and model the energy use of buildings. Executives and line-of-business managers use the data to determine potential energy generation from rooftop solar panels. Designers of renewable energy systems use it to size solar panels and plan energy backup systems. This data also helps farmers and commodity traders calculate expected crop yields, because solar radiation is a key input to crop growth. And NASA even sees data downloads along shipping routes, for ships equipped with wind and solar power systems.

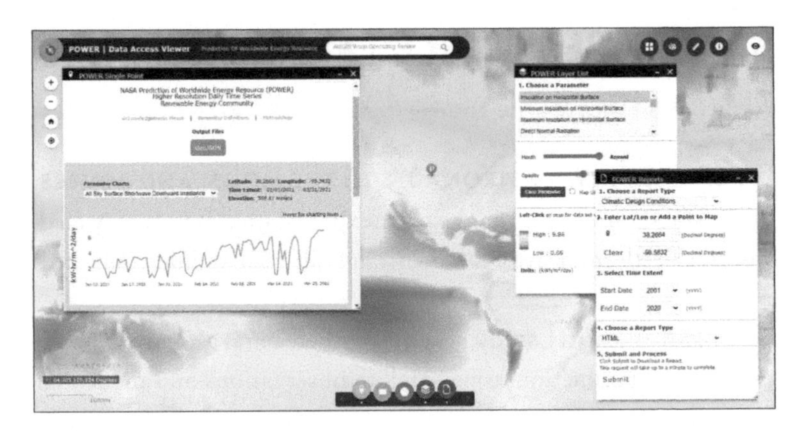

The POWER team provides the Data Access Viewer, a GIS web-based tool that allows anyone to see and explore a variety of renewable energy variables at any location around the world.

Since POWER adopted GIS capabilities, the team has fielded millions of requests from hundreds of thousands of users.

Earth observations for renewable energy

Since the POWER project began in 2002 with the goal of making surface energy measurements available, it has seen tremendous growth in the data it aggregates and provides, the answers it can offer, and the communities it serves.

In the transition to clean energy, the POWER project acts as an accelerant, providing time series and climatological data as a service, which helps inform planners of what to expect in the future. Satellite observations combined with sophisticated computer models identify solar and wind energy resource availability around the globe. For less developed areas or places without on-the-ground data or a long history of it, derived observations filled in by a model can provide accurate estimates.

The POWER team provides the Data Access Viewer, a GIS

web-based tool that allows anyone to see and explore a variety of renewable energy variables at any location around the world. Software developers can connect directly to the application programming interface (API) and use the data within the applications they build, and cartographers or other GIS professionals can use ArcGIS image services to perform analysis and create smart maps and dashboards that support renewable energy plans.

For example, the local government of Satellite Beach, Florida, in collaboration with NASA, used POWER data to estimate the role rooftop solar can play in the city's goal of generating 100 percent of its energy needs from renewable sources by 2050. In the equatorial region of West Africa, a researcher attributed a decrease in solar panel performance to cloud cover. Instead of blaming solar panel performance, the community added more panels to achieve its energy needs.

How maps connect data to deepen understanding

Decision-makers working on renewable energy initiatives rely on shared smart maps for a bird's-eye view of on-the-ground opportunities. Deciding where to generate power from renewable energy requires answers to a series of location-based questions that start with where resources are likely to be the greatest, followed by considerations of local demand and the cost of connecting to the grid.

Spatial analysis makes it easier to answer these questions. Within GIS, a smart map can layer solar and meteorological data from NASA with socioeconomic and energy grid data. In this way, modern GIS software helps make sense of massive volumes and varieties of data, showing an accurate picture of current conditions and empowering users to model and simulate strategies.

With growing and widespread demand for renewable energy, the detail amassed from GIS and smart maps also allows decision-makers

to compare and prioritize opportunities at multiple locations. NASA's observations are freely available to everyone.

A version of this story by Patricia Cummens originally titled "Mapping Renewable Energy Potential with Help from NASA POWER" appeared in the *Esri Blog* on March 22, 2022.

HOW AND WHY UTILITIES ARE SHIFTING TO RENEWABLES

Xcel Energy, Austin Energy, and Ørsted

IN 2018, MINNEAPOLIS-BASED XCEL ENERGY BECAME THE first major US utility to commit to being carbon-free by 2050. Shortly afterward, Xcel competitor Platte River Power Authority, located in Colorado, announced it would end carbon emissions by 2030. In the wake of these moves, a wave of smaller utilities followed with plans to decarbonize on even shorter timelines.

It was a sign that for many operators, clean energy is no longer just an environmental or political prerogative; the incentive to decarbonize has become an economic one. The market is reaching a tipping point where wind and sun power are cheaper than fossil fuels. Decentralization, demographic changes, and digitization are accelerating the transition.

Utilities are shifting toward a more customer-centric model, built on data about where the demand is and where it will be in the future. This shift requires technologies that can provide real-time, highly detailed consumer updates. Increasingly, utilities are turning to GIS to guide location-based decisions about everything from managing load on the grid to siting renewable energy infrastructure such as wind turbines and solar panels.

Another relevant application of GIS has been its ability to provide deeply nuanced portraits of customers based on neighborhood, profession, age, income, and other demographic variables. These perspectives recognize the importance of consumer choice and the increasing diversity of energy consumers.

A new, customer-focused model of energy

Utilities once viewed customers as more or less anonymous ratepayers, broadly classified as residential, commercial, or industrial. Today, utility executives require a more detailed understanding of consumers, addressing questions about who is most likely to buy an electric vehicle (EV) or install solar panels, what neighborhoods will need charging stations, and where to offer incentive programs to manage load.

GIS mapping applications can model customers and their usage habits with geographic accuracy to help answer these questions. Environmentally focused groups, "prosumers" (adapters of green energy who themselves supply power), and other demographic trends make incorporating GIS-based location intelligence into a business strategy even more advantageous. Utilities are now gauging how quickly and smartly they can transition to clean energy.

Clean energy's economic tipping point

As the costs of green energy drop while its effectiveness increases, the economic argument for decarbonization has become difficult to ignore. According to one estimate, the total capacity of solar and wind generation in the United States grew from 9.5 gigawatts in 2005 to 113 in 2017.

A study by renewables analysis firm Energy Innovation found that around three-quarters of US coal production is costlier than wind and solar energy in supplying households with electricity. Accounting for subsidies, the megawatt-hour cost of electricity from wind power had fallen as low as $14 by 2019, according to investment firm Lazard's analysis. By comparison, running existing coal plants cost an average of $36 per megawatt-hour, while the cost of creating and operating new coal plants was as high as $143.

Consumer choices reflect the new reality of green energy. In

2017, 3.1 million EVs were in use worldwide—a 54 percent jump from the year before—according to the International Energy Agency's Global Electric Vehicle Outlook 2018.

Utilities' road map to innovation

In the midst of these changes, utilities occupy a paradoxical position. As highly regulated, vertically integrated monopolies, they tend to be risk-averse and slow-moving. At the same time, they are attuned to future trends.

Kevin Prouty, group VP of energy and manufacturing insights at International Data Corporation (IDC), said that utilities "are probably the most forward-thinking industry in the world. They're looking 20, 30 years out." They can see the changes to the industry coming, even if they can't react as fast as private companies might.

Utilities are also magnets for data, often gathering it faster than they can process or analyze it. Imagine a typical large utility with 2 million meters, Prouty suggested. If they're collecting data every 15 minutes, that's 8 million data points every hour for 24 hours a day.

Utilities must make sense of all this data in a way that yields insight, ideas for best practices, and forecasts that align with long-term strategy. That's where GIS comes into play.

"I think everything a utility does is around location," Prouty said. "That's why, if you look at the utility space, every utility has some form of a geospatial system. They have to."

Austin energy drives EV revolution

Lindsey McDougall, an EV program manager at Austin Energy in Austin, Texas, can remember a time when she had visited every single EV charging station in the city. "I can't do that anymore," she says. "There's just too many."

Since 2013, McDougall has overseen operations for EV adoption, with a focus on incentive programs that offer rebates to EV drivers as they use charging stations around town. Six years after these programs were put into place, Austin Energy experienced growing pains as adoption levels for EVs multiplied. "They're coming on our network so fast," she said.

To help deal with the tide of customers and prepare for a future focused less on incentives and more on demand response, McDougall relies on GIS and the location intelligence it supplies about customers and their habits to prepare for the future.

In particular, McDougall uses GIS-driven demographic analysis to drill down into the specifics of where demand is coming from and where it might increase. By classifying neighborhoods into dozens of consumer segments organized by postal code and socioeconomic data, GIS can provide utilities with insight into residents' lifestyle choices that relate to EVs and other forms of green energy.

By comparing the profiles of existing EV drivers to the entire customer base, Austin Energy can use targeted marketing to reach out to prospective customers about the EV programs in enticing ways.

"We can really classify them by address," McDougall said. "Through the [demographic] data, we see how they prefer to be reached out to, via mail or email or phone."

GIS has also had broader applications for Austin Energy in how it cooperates with Austin city political leaders to meet transportation and sustainability goals. When the Austin City Council recently revisited its *Imagine Austin* road map for sustainability, McDougall used smart maps to explore ways to incentivize deployment of charging station infrastructure that will create a more connected city.

When Austin Energy's project leader presented to company executives on the deployment of DC charging stations, they did so with one of McDougall's maps in hand. That map emphasized the wide geographic distribution of the DC stations—an important feature in

Austin, where council members want to ensure that all communities, especially underserved ones, are being addressed.

"GIS is used for a lot of visualizations to help us really understand what's going on," McDougall said.

Green energy changes in Denmark

The shift toward renewable energy sources and the need for map-based technology to help utilities accommodate it is underway in Europe. Ørsted, based in Denmark, has become one of the world's largest offshore wind energy developers—a remarkable turnaround for what was just years ago a foundering energy company focused on coal-based power plants and petroleum.

Ørsted sold its oil and gas holdings in 2017 and increased its investment in offshore wind, which moved the organization to the global forefront of green energy. Denmark company competes with prominent multinational companies for international markets, with its wind farms already established throughout northern Europe. It also planned an offshore energy project in New Jersey to power a half-million homes.

Making this shift has put Ørsted in the position of managing a broad base of customers and facilities. Jakob Mortensen, former head of energy analytics and geo services at Ørsted, was at the center of some of these changes. Although Ørsted was an early adopter of GIS, employing the technology more than 20 years ago, Mortensen said the demand for location intelligence has grown, leading the company to require more detailed levels of data.

"We have to look at our customers in a completely new way compared to only a few years ago," Mortensen said. "Usually, people had to buy all their electricity from one company, but now they have an option to choose their own vendor. At the same time, we also see a huge increase in the amount of solar panels on private households.

We expect to see a huge increase in privately owned and company-owned batteries."

Ørsted relies on GIS as a tool to illuminate demographic change, predict where solar panels will become more popular, and determine how to apply the current potential of grid capacity.

Denmark limits the amount of information that utilities can acquire about individual consumers. But even by paying attention to location intelligence on a regional level, they can identify and forecast trends. Areas with dense concentrations of academics with high incomes, for example, signal a demographic that is more likely to purchase EVs.

The other major area where GIS has proven itself for Ørsted is in guiding the placement of wind farm turbines. Industries such as internet service providers are similarly relying on GIS to branch out into new regions. As Ørsted seeks to find good locations for offshore wind farms and where to lay down cables, location intelligence can also collate data relating to bird migration patterns, shipping lanes, and unexploded bombs from World War II to aid placement. As Ørsted's presence expands internationally, reliance on GIS will come into even greater play.

For Mortensen, the central argument for using GIS comes down to a financial one. He recalls a quote that came up at an industry conference a few years back: "Why spend billions in copper when you can use millions in software?"

Having data with a geographic awareness means greater efficiency, less waste, and a more confident path into the future. "We can save a lot of money by knowing more about how our grid is configured and how it's loaded," Mortensen said.

Putting consumer choice first

As some of the most consequential players in the shift to clean energy, utilities are poised like few others to employ location intelligence to prepare the grid for the future.

The relevance of GIS to new infrastructure is clear, but how GIS can connect utilities to customers may prove even more essential over time. In a world where consumers have more choices in their energy sources, a deeper understanding of who the customer is and what they want will be a utility's defining competitive edge.

A version of this story by Matt Piper originally appeared in *WhereNext* on August 27, 2019.

MAPPING THE FUTURE OF ENERGY THROUGH A GEOGRAPHIC APPROACH

Energy Queensland

ENERGY QUEENSLAND SUPPLIES ELECTRICITY TO AN AREA of northeastern Australia four times the size of California. It operates 130,000 miles of power lines (enough to stretch around the world five times) and 153,000 substations. It's one of the largest energy networks in the world, scattered across tough and mostly rural terrain.

Maintaining and upgrading that network has always meant dealing with vast, inhospitable geography and extreme nature. Every year Queensland residents have to deal with more extreme bushfires, cyclones, and floods, attributed at least partly to climate change.

As a result, providing the state with stable and secure, low-cost energy is an increasing challenge for Energy Queensland. But the government-owned provider has greater ambitions and is set on transitioning to 50 percent renewable energy sources by 2030. In scale and speed, it's an exceedingly ambitious target—one that other energy companies around the world are watching and following.

Mapping energy transition

The company now distributes energy from thermal, hydro, wind, and solar power sources, including 37 industrial-sized solar power farms. But over a third of its solar power comes from customers and domestic solar panels, and that is likely to rise to one half over the next two or three years. Feeding that varied and disparate supply into the network is challenging, as is dealing with peaks and troughs in demand as more of its customers generate their own power at least some of the time.

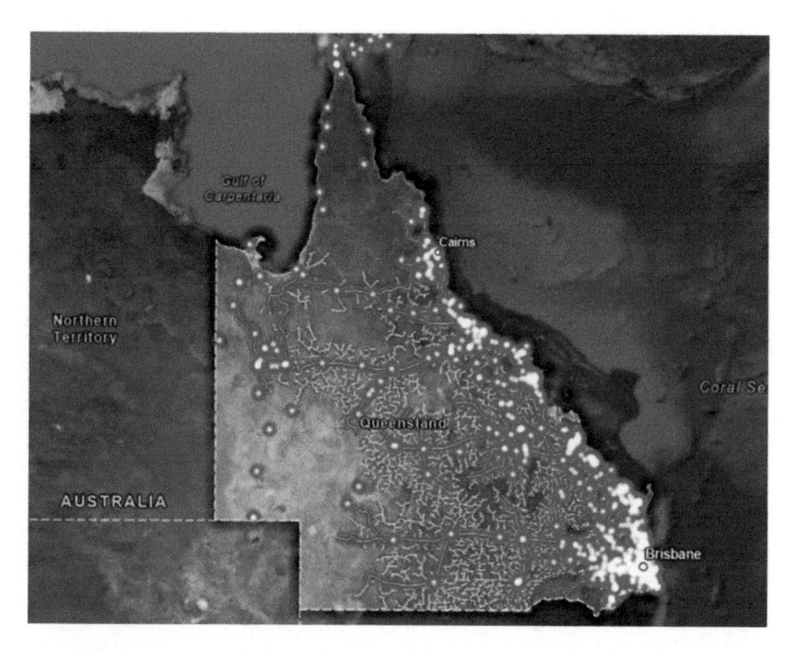

Energy Queensland infrastructure map. Source: Energy Queensland.

Energy Queensland could not navigate these challenges without an exact mapping of its assets and how those assets connect. The utility used GIS to create a 3D digital twin of its entire network, above and below ground.

"Because location was so critical, we corrected all of the asset locations from our old dataset to real-world positioning," said Shannon Connolly, manager of spatial enablement at Energy Queensland. Creating that digital twin means it can add layers of dynamic information, taking in landscape and heritage sites, weather, and shifts in supply and demand.

The company's first challenge is resilience of the network in the face of more extreme weather. GIS improves predicting the impact of those events and the response time for repairing outages. In addition, customers can be kept better informed about outages and repairs.

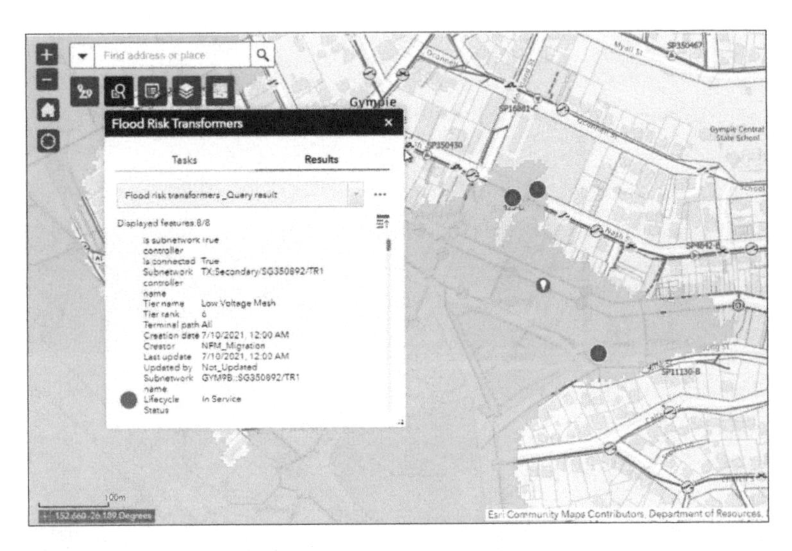

Map of electric transformers at risk from flooding. Source: Energy Queensland.

This level of location intelligence is also vital in managing the rapid transition to renewable energy. "One in three houses in Queensland now has solar power stations on its roof that supply the house but can also put power back into the network," Connolly said. "I don't think people actually realize the complexity that moving to solar and sustainable input introduces to the management of the actual network and maintaining a reliable secure supply."

As Connolly said, Energy Queensland's GIS-backed modernization is a valuable test case. "We have always thought of this as a transformational project, not an IT project. Its size and complexity have triggered interest across the industry. Over the last year and a half, we have been contacted by a dozen major utilities in Australia who are looking to make the same transformation. They all understand that the existing business model doesn't support demand or customer expectations."

Energy Queensland is taking an open-source approach, sharing its modeling and business principles to "assist this uplift in the entire Australian energy industry." They are now duplicating this process with several utilities across the United States, Canada, and Indonesia.

"As we have found, it doesn't matter where the industries are, they always have the same issues: weather events, maintenance, dealing with demand, the switch to sustainable energy and how to manage that, and how to record all that and show it in your systems. So when I say it's transformational, it's not only transformation for Energy Queensland; we're hoping it's a trigger for transformation of our industry as a whole."

A version of this story originally appeared in *Economist Impact*.

PART 3

ENVIRONMENTAL MONITORING

BALANCING THE QUALITY OF LIFE WITH EQUITABLE economic and environmental sustainability requires careful monitoring. To address ecological challenges, government regulators use GIS to monitor the presence of harmful contaminants and pollutants and identify and resolve environmental threats. ArcGIS provides in-depth location- and science-based insights that present unbiased environmental monitoring, assessment, and response solutions and policy to protect lives and the environment.

Equitable environmental monitoring starts with GIS

A robust foundation for equitable environmental monitoring starts with understanding environmental impacts, monitoring, and assessment. The ArcGIS geospatial infrastructure helps users connect data, people, and predictive analysis, providing an interdisciplinary approach to equitable environmental monitoring:

- **Get science-based insights:** A location-enabled, common operating picture unifies data to allow for real-time situational awareness and optimal environmental monitoring.

- **Explore environmental conditions:** Analytics and visualizations show how conditions have changed over time and how they impact the human population.
- **Receive inclusive assessments:** Geospatial tools help streamline equitable environmental assessment to understand the environmental impacts of proposed projects, ensure course corrections, and achieve permit compliance.
- **Transform mobile field collection:** These tools also convert paper-based inspection to digital forms. Users can enable one system of record for streamlined field data capture and sharing to the office, with automated reports and alerts for faster response times.
- **Modernize site inspections:** Using ArcGIS, regulators can collect data, conduct inspections, prepare regulatory reports, and align the information to ensure regulatory compliance.
- **Create collaborative plans:** GIS delivers a collaborative framework for iterative design scenarios informed through modeling and endorsed through stakeholder engagement to support optimal outcomes.

GIS in action

Next, we'll look at some real-life stories of how organizations are using GIS to increase environmental sustainability and ensure equitable environmental monitoring.

USING MAPS TO REVEAL INJUSTICES

US Environmental Protection Agency

WHEN CHARLES LEE FIRST EXPLORED THE TOPIC OF environmental justice in the early 1980s, he was in uncharted territory. "The whole notion that there are certain people who are disproportionately affected by environmental harms was something that nobody thought about," said Lee, currently a senior adviser for environmental justice at the US Environmental Protection Agency (EPA). "It was prevalent, but nobody knew what to call it."

One reason for the disconnect was that modern environmentalism, which gained force in the 1960s, was mostly committed to issues around wildlife and conservation, with little attention paid to issues of equity and social justice.

At the same time, the modern civil rights movement focused on racial disparities in education, employment, and economic advancement. And so, throughout the 1970s, the two movements occupied mostly separate political universes.

Environmentalism and civil rights

Today, experts have repeatedly established the link between environmental and equity issues. Communities of color disproportionately bear the consequences of excess pollutants and contaminated soil and water, including higher-than-average rates of diseases ranging from asthma and cardiovascular illness to cancer, birth defects, and other health disorders. Scientists, policy makers, and activists often bolster their findings by using GIS technology, creating interactive fact-based maps to bring the issue into stark relief.

Early in 2021, activists in Chicago staged a 28-day hunger strike to protest the proposal to relocate a metal-scrapping facility to the city's South Side. The industrial process, they pointed out, creates the kind of air particles that cause health problems such as asthma.

Pumps draw petroleum from oil wells near homes in Los Angeles.

The activism in Chicago recalls the seminal moment in the history of environmental justice that galvanized people such as Lee. In 1982, protests erupted in Warren County, North Carolina, over a plan, supported by the state's governor, to dump thousands of tons of contaminated soil in a low-income, mostly Black neighborhood.

For six weeks, the protests made international news. Although the demonstrations were unsuccessful in blocking the plan, they represented a critical union of environmentalism and civil rights advocacy, helping create the modern environmental justice movement.

Toxic waste: A national problem

In the aftermath of the Warren County protests, the US Government Accountability Office (GAO) examined the demographics of hazardous-waste sites in the southeastern part of the country. The subsequent study revealed that three out of four sites were in Black-majority counties.

When the study was released, Lee was directing a project on environmental hazards in areas where low-income communities and people of color lived. Lee was working with the United Church of

Christ's Commission for Racial Justice, which had played an active role in the Warren County protests. The GAO study confirmed for him the widespread problem linking disempowered communities and hazardous waste sites.

"I was discovering more examples of how communities of color—from urban areas to reservations to the farmworkers—were affected by environmental hazards," he said. "I realized that if you could replicate the GAO study on a national level, it could really put the issue on the map."

Mapping evidence of environmental racism

For several years, Lee gathered data and used maps to illuminate racial disparities. The landmark report, *Toxic Wastes and Race in the United States*, published in 1987, became the founding document of the environmental justice movement.

Working from a definition of racism as being "racial prejudice backed by power," Lee's team revealed, with extensive empirical evidence, the power at stake. The report quoted an EPA estimate that the total cost of ameliorating the country's thousands of uncontrolled toxic-waste sites could exceed $100 billion (adjusted for inflation, over 1 percent of the US gross domestic product [GDP] in 2021).

Lee's team found that three out of five Black and Hispanic Americans lived in communities near these sites, as did 2 million residents of Asian and Pacific Island heritage and 700,000 Native American communities. This expensive problem showed up in the communities least equipped with the political and economic tools to demand solutions.

Lee's team also discovered a strong correlation between the percentage of people of color in a community and the amount or volume of active hazardous-waste facilities. Race proved to be the most significant factor in predicting the location of the facilities.

"The possibility that these patterns resulted by chance," Lee wrote, "is virtually impossible."

"Something you can't contest"

Environmental justice is fundamentally a matter of race and place, making geography fundamental to the solution. Practitioners must examine the connection between communities and their environments—the land people live on and the air they breathe. Maps turned out to be the most potent way to demonstrate these connections.

The rise of the environmental justice movement has coincided with the increase in capabilities and accessibility of GIS software that can process, aggregate, and display these connections on smart maps.

"The maps in *Toxic Wastes and Race* were all hand drawn," Lee said. "GIS existed, but it was way beyond our means to get access to it."

As GIS has become more commonplace, it has also become integral to thousands of environmental justice projects. It supports arguments while also making evidence easier to communicate.

"If you talk to people who do community organizing with [GIS], they say what makes it so powerful is that it puts the issue in front of people as something you can't contest," he explained. "It gets people excited about talking about these issues because it makes it so real."

Tai Lung, a geographer who oversees the EPA's environmental justice mapping and screening tools, agreed. "Being able to put GIS in the hands of people really democratizes the information," said Lung, whose work was noted in the Justice40 Initiative, a 2021 Biden administration directive to address decades of underinvestment in disadvantaged communities affected by climate change, pollution, and other environmental hazards. "There are communities that have lived in polluted conditions for decades, and they've often been ignored. But now they can come back and say 'Look, we have all this data that we can pull from this really handy tool.'"

The EPA's EJSCREEN online mapping tool lets users explore minority and low-income populations and potential environmental quality issues.

Critical challenges ahead

Environmental justice experts also found GIS useful in providing a temporal aspect to data, illuminating the historical entrenchment of equity-related problems. Lee pointed to the Mapping Inequality project developed by the University of Richmond, Johns Hopkins University, and Virginia Tech University in 2016. The project uses real estate maps from the 1930s and 1940s to show how the neglect of certain neighborhoods is rooted in discriminatory real estate practices collectively known as redlining.

During the coronavirus pandemic, epidemiologists used GIS to understand the spread of COVID-19, and environmental justice advocates used the technology to communicate related response disparities.

This innovation may show how GIS can most help the environmental justice movement by underlining the unnatural aspect of the impact of natural disasters.

"Natural disasters are always going to happen," Lung said. "But who's going to be impacted the worst? The people in the low-lying areas of New Orleans will get flooded—not the people in the Garden

District. And mapping gives you a way to layer on demographic information and say, 'This is where the Black folks live.' You can layer it all together and see direct correlation between all the data."

Lee echoed Lung's assertion and underscored how maps continue to support the mission.

"Environmental justice is proving to be an increasingly powerful lens to understand a lot of the critical challenges of the 21st century," Lee said. "GIS can help shape the conversation in ways that engage people, helping them to identify solutions."

A version of this story by Margot Bordne and Clinton Johnson originally titled "Charles Lee: Environmental Justice Leader Uses Maps to Reveal Injustices" appeared in the *Esri Blog* on June 30, 2021.

INTEGRATING EMERGENCY RESPONSE

US Environmental Protection Agency and Tetra Tech

THE EPA PROTECTS HUMAN HEALTH AND THE ENVIRONMENT. Its Office of Emergency Management collaborates with federal partners to prevent accidents and maintain well-tuned incident response capabilities. Following Hurricane Irma in August 2017, the EPA needed to facilitate communication among stakeholders from local, state, and federal agencies. The hurricane had unleashed contamination sources, such as compromised facilities and containers, across a wide area. The EPA wanted to create and share a single common view of these environmental threats so that stakeholders could rapidly identify and assess response targets and prioritize cleanup efforts. They could then plan appropriate mitigation efforts to lessen or prevent environmental impacts.

Tetra Tech, an Esri partner, provides consulting and engineering services. The company supports government and commercial clients with solutions focused on water, environment, infrastructure, resource management, energy, and international development. Tetra Tech has worked with the EPA on contracts across the United States. Throughout the EPA's Hurricane Irma response, Tetra Tech provided technical support for collecting, tracking, analyzing, and communicating the status of response targets. The EPA needed to integrate information from sources and synthesize it into a common and updated view for all stakeholders. Tetra Tech configured a focused solution built around ArcGIS Online.

The EPA and Tetra Tech built an emergency response solution using GIS technology, which generated a common operating picture (COP). Using ArcGIS apps, the EPA digitally captured field data. Staff used ArcGIS Collector to edit, update, and create features that represented response targets. Ground crews used ArcGIS Survey123

EPA staff use Collector to edit, update, and create features that represented response targets, and ground crews use it to gather form-based information and transmit it to the emergency operations center.

to gather form-based information and transmit it to the emergency operations center. ArcGIS Workforce helped response teams assign, manage, and track the status of tasks such as assessment, identification, and mitigation. To automate tasks and manage data, staff relied on ArcGIS API for Python and ArcGIS Pro, which ensured that incoming data sources were integrated and synced to operational data.

Stakeholders used ArcGIS to access, edit, and contribute operational information from the field or office—whether they were connected to the internet or not. By drawing technology components into one integrated system, ArcGIS reduced responder redundancy because everyone was working from the same map. Furthermore, the solution performed low-value activities such as data processing so that the EPA could focus on high-value activities such as analysis and data-driven decision-making. Feeding real-time information into the COP, the emergency response solution heightened stakeholders' operational awareness so that they could immediately follow unfolding events.

A version of this story originally titled "EPA Integrates Emergency Response" appeared on esri.com.

MAPPING EQUITABLE DISTRIBUTION OF INFRASTRUCTURE FUNDS

Montana Department of Natural Resources and Conservation

WHEN THE MONTANA DEPARTMENT OF NATURAL Resources and Conservation (DNRC) found itself with a rare budget surplus, state officials analyzed maps and built online dashboards to share results with the public.

The $900 million surplus resulted from the $350 billion American Rescue Plan Act of 2021 (ARPA) passed by Congress to build the economy at the local, county, state, and tribal government levels.

The Montana legislature decided to spend most of the money on upgrading water and sewer systems and directed the DNRC to determine which communities to invest in.

The DNRC invited city, county, and tribal governments; state agencies; water and sewer associations; and conservation districts to submit project proposals to fund. "They only had about a month to apply for this money, and we got over $900 million in grant requests," said Autumn Coleman, DNRC's resource development bureau chief.

To allocate and disperse the funds, DNRC created two separate grant programs. One divided the grants among Montana's 56 counties, based on the same formula used to distribute money collected from the state's gasoline tax. This gas tax formula calculates a county's size, population, and miles of public roadways. The formula was fair but complex and difficult to parse for those unfamiliar with it. Adding further confusion, potential grantees had to present matching funds.

That's where an online dashboard and map supported the effort. Coleman and her team reduced the complexity of the process by building a public-facing map and dashboard using GIS.

The public can view the dashboard of fiscal recovery fund spending on water and sewer infrastructure.

The visualization tools added a graphic element to what were otherwise facts and figures. Residents and grant applicants could more easily connect the gas tax formula to its impact on funding decisions—and also monitor what decisions were made as they began to appear on the map.

"It's just a really hard thing to explain to folks, so having this mapping tool made it easier for people to understand," Coleman said. "We could share a giant 20-page table, but it just doesn't play as well as being able to look and see what you and your neighbors are getting."

One big state, many small identities

The second ARPA grant program was competitive. Proposals were judged solely on merit. Geographic distribution was not officially considered, but it wasn't ignored either.

Like all states, Montana has important political and demographic considerations, which are influenced by its size as the third-largest continental state in the country but sixth lowest in population. Natural conflicts included urban centers versus rural areas, sparsely

Montana is well known for its water resources, such as Flathead Lake, which is the largest freshwater lake west of the Missouri River.

populated eastern Montana versus the more populous west, and longtime residents of modest means versus more affluent new arrivals. This all played out against the backdrop of Montana's rapid population growth, enough for the 2020 census results to earn the state a new congressional seat. The fastest-growing area is Gallatin County, one of only two Montana counties with a median family income greater than the US average.

Despite its growth and reputation as a beautiful place to live, Montana was ranked as the nation's seventh-worst state for infrastructure. According to Coleman, water and sewer funding is viewed as an important dimension of fairness by the state's differing populations and places and their representatives.

"We're a very dispersed state, with a few major population centers and lots of rural areas," she said. "Legislators want to see ARPA funds spread evenly across Montana. They want to make sure that the smaller communities in their districts get a fair chance in all of this."

Healthy competition

DNRC's map and dashboard of competitive grant applicants served a dual purpose. The map and dashboard promoted transparency by showing the public the geographic distribution of the grants and allowed Coleman's team to test funding scenarios, which in turn supported internal accountability.

"It was definitely a decision-making tool," Coleman said. "We could look at it and say, 'If we fund the top 28 projects, what does the map look like?' As we funded more projects, we could see the dots spread across the state."

The dashboard and maps showed the distribution of funds beyond population centers and areas disproportionally affected by COVID-19. For Brian Collins, GIS manager at DNRC, the maps served as a manifestation of the agency's larger purpose—serving residents.

"Putting together this kind of information resource was a good reminder that we're in a public service profession," he said. "We're providing customer service at a very high level to people that need it right now. And it's very gratifying to put it out there this way."

A version of this story by Christopher Thomas originally titled "Maps Help Ensure Equitable Distribution of Infrastructure Funds in Montana" appeared in the *Esri Blog* on November 9, 2021.

GLOBAL COLLABORATION FUELS ENVIRONMENTAL IMPACT ASSESSMENT

Bioinsight

CONSULTING WITH COMPANIES ON SUSTAINABILITY practices and environmental compliance is a growing business, projected to become a $36 billion slice of the global economy by 2027.

That growth is fueled in part by increasing consumer interest in climate-friendly practices and a corporate trend toward improved environmental stewardship. Because of those forces and the ongoing drive for regulatory compliance, it's vital for companies to receive fast, accurate environmental impact assessments so they can continue to grow responsibly.

In meeting this need, one environmental consulting firm headquartered in Lisbon, Portugal, discovered the value of GIS. Company leaders knew that GIS technology could analyze massive amounts of data and plot the results on smart maps. In using GIS, the team at Bioinsight found that maps are also tools for collaboration, communication, and project management.

Through mobile devices and a new generation of lightweight GIS apps, Bioinsight project managers stay in touch with the work of biological survey teams in remote areas on several continents.

As the environmental impact assessment industry grows more competitive, Bioinsight relies on location technology to help clients achieve their sustainability goals and complete projects faster and more accurately.

Location intelligence for environmental impact assessments

As an environmental consultancy, Bioinsight analyzes how planned wind farms, pipelines, power substations, roads, and other structures will affect animal and plant life in areas under consideration for development.

Its environmental impact assessments give companies the data they need to lessen projects' ecological impacts, ensure regulatory compliance, and address any additional concerns.

Bioinsight CEO Miguel Mascarenhas said it's important to perform early location analysis and deliver quick, efficient impact assessments to help companies avoid environmental problems that might otherwise emerge after a site is operational.

For example, the company will assess corridors for a road before one is planned to see which option has the least environmental impact and which corridors to avoid, he said.

The road might provide access to a wind farm, the most frequent subject of Bioinsight's environmental impact assessments. These sustainable energy projects continue to gain favor with customers and energy providers around the world, but they can't be built without impacting the surrounding location.

Bioinsight documents and digitizes the types of birds and their flight paths near proposed wind farm sites. Using mobile apps to do this work saves time by automatically capturing the locations.

As competitors enter the marketplace, Bioinsight's use of GIS for communication and project management has been a game changer in terms of competition, according to Mascarenhas. Bioinsight's teams routinely use GIS mobile and desktop apps to monitor progress on different continents.

"If you have a plan where we need to go to 30 or 40 different places to collect data in remote areas, sometimes access is not easy, and we don't have any type of good reference points," Mascarenhas said. A smart map can identify which tasks and information were completed on a given day.

Improved collaboration saves time, money

A few years ago, the process of sending teams of biologists into remote areas to survey animal and plant species could be costly, time-consuming, and frustrating.

On-site researchers logged the type, number, and movement of species on paper along with details such as map coordinates, day, date, time, temperature, and weather. Researchers mailed that information and other project data to Bioinsight headquarters in Lisbon, where others entered the information into spreadsheets for analysis.

The older method took as many as five days to send collected data through the mail and update a digitized report. It increased the handling of information, resulting in more mistakes.

If the analysis revealed that the researchers were in the wrong spot or not observing the specified plants or animals in the required

Biologists in the field use smart maps on their mobile devices to show where they need to survey animal and plant species and log sightings.

numbers, the project manager might need to direct the team back to the remote area to gather necessary data, a costly consequence in terms of time and money.

Bioinsight staff found they could communicate more efficiently through GIS technology.

"While the team is in the field, I can start looking at the data and make sure that the data are accurate and are the data that I needed," Mascarenhas says. "If not, I can contact them and say, 'I need you to spend another day on the field.' That's important for us because sometimes I'm here in Portugal, but our project is in Brazil or is in South Africa."

Communication with smart maps

When biologists in Brazil or elsewhere have smart maps on their mobile devices, data is more accurate, collaboration is more natural, projects run more efficiently, and costs decline.

Anyone in the company who needs to review an environmental assessment can check the shared smart map, a process that often alleviates the need for email or phone calls.

Even if team members must wait until they return to their lodgings at night to connect to the internet, the latest data is available a few hours after it has been collected—much more efficient than the multiday lags that used to occur.

Expanding to additional stakeholders

Bioinsight's collaboration doesn't stop with company employees or subcontractors. The firm has found ways to extend the collaborative capabilities of GIS to its work with local residents. This allows more stakeholders to get involved in high-interest projects by performing tasks that don't require trained biologists.

In the northern part of Portugal, Bioinsight's staff set up cameras

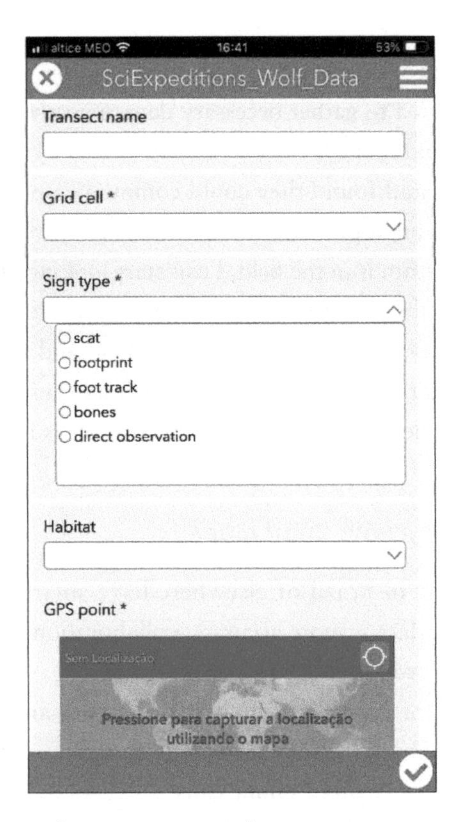

GIS data collection for an environmental impact survey.

in about 30 locations to document movements of the wolf population—an endangered species in Portugal and one that sometimes preys on livestock or pets. The goal was to document the movement of the small wolf population, educate residents, and promote conservation.

"It's a collaboration between us and the local people to join together in a project with a species of wolf that's sometimes problematic," Costa says. "It's one of the examples where we can use GIS."

Bioinsight scientists set up and then relied on a volunteer team to collect data.

The mobile smart maps guide volunteers to the right locations. Once they've found the cameras, they send the SD cards to Lisbon. There the Bioinsight team analyzes the images to understand wolf activity and monitor the health of the local population.

"I don't have to go on a second trip, spending the expense of going there," Mascarenhas says. "It's a good example also of how we can save money and interact with a local team."

Where next? More collaboration, greater speed

The head of Bioinsight anticipated increasing use of GIS and smart mapping for communication and project management going forward.

"It's data communication and speed," Mascarenhas said, "to have the data as soon as possible on our desktop so that we can start analyzing and making decisions regarding the subject we are studying."

A version of this story by Audrey Lamb originally titled "Environmental Consultant Achieves Global Collaboration with Smart Maps" appeared in *WhereNext* on June 29, 2021.

3D MAPPING HELPS PRESERVE FRESHWATER RESOURCES

US Environmental Protection Agency

THE EPA, WHICH SAFEGUARDS WATER QUALITY AND determines the use of water resources, continues to replace paper-based maps and aerial photography prints with GIS and imagery.

The demands of climate change and population growth affect lakes, rivers, streams, and underground aquifers that supply fresh water. Across the United States, communities of all sizes face the need for increased, sustainable access to freshwater resources.

Although analog methods served EPA in the past, they are no longer efficient, and the agency replaced them with digital ones. The EPA employs geospatial tools to complement its standard photogrammetry practices. Photogrammetry is the science of obtaining reliable measurements from photographs and digital imagery to produce

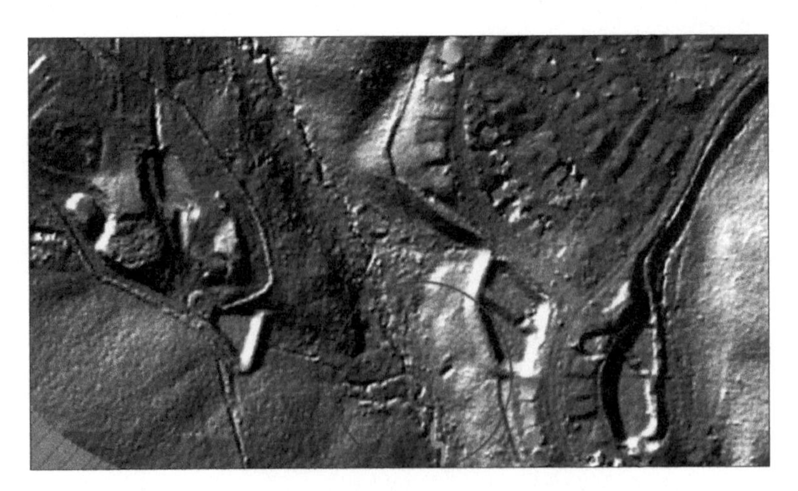

USGS stream data (blue line) is compared with stream channel characteristics expressed by hillshade elevation data derived from lidar.

products such as orthomosaic maps, GIS data layers, or 3D models of real-world objects.

The EPA established a process to develop information that supports enforcement of its cases. Once the agency embraced remote sensing and digital mapping, it realized that these tools could help with its investigations. GIS helps answer questions about how wetlands have changed over time and understand the impacts of its decisions on commerce.

Bringing orthomapping and digital maps together

The EPA has begun using orthomapping in ArcGIS Pro. More advanced image analysis capabilities, provided by the Esri orthomapping suite, are available in ArcGIS Image Analyst.

With orthomapping, historical information augments current data to make better decisions that support the EPA's work enforcing federal environmental laws that provide funding to clean up uncontrolled or abandoned hazardous-waste sites.

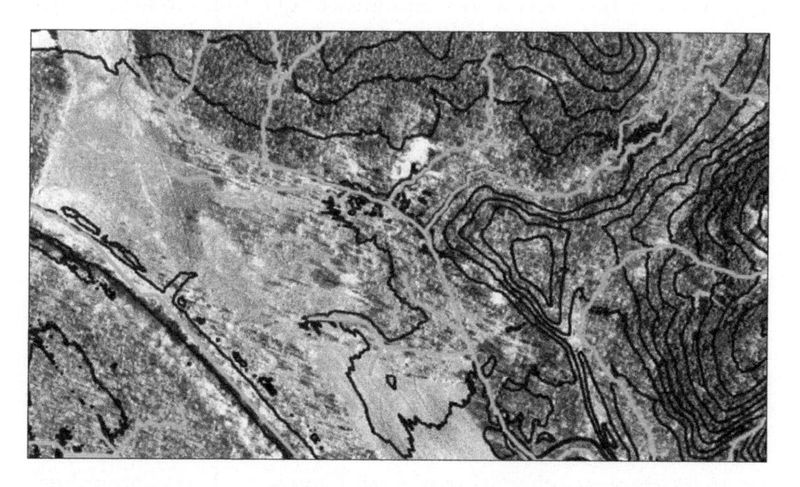

Contours and stream network data that was derived from digital elevation data using ArcGIS Pro are overlaid on historical aerial photography.

Orthomapping capabilities in ArcGIS Pro support EPA studies by accurately placing raw historical photos into the same view with existing datasets and maps. Before transitioning to ArcGIS Pro, the EPA viewed paper maps and photographic transparencies on a light table with traditional optical devices. It was a slow and tedious process that involved manually positioning film or prints that could easily get out of alignment.

Changes over time

Peter Stokely has worked as an environmental scientist at EPA for more than 30 years. His work centered on using imagery analysis and GIS to provide accurate analyses of water resource sites. He has focused on compiling evidence to support enforcement of the Clean Water Act, which protects wetlands, streams, and rivers from being filled with solid material.

Stokely uses GIS and photogrammetry to support advanced spatial analytics. He looks at current and historical imagery to digitize changes over time. His work helps answer questions about the location of wetlands under federal authority and whether those lands are managed in compliance with the law.

Historical imagery is useful for more than understanding landscapes. It helps in evaluating the historical impacts to a local water resource. This imagery is necessary when studying a property that is subject to pollution to determine the source and timing of contamination and provides the EPA with context for enforcement. Asking questions about an area's past can help guide decisions and identify potential future implications.

"You can make a lot of interpretive mistakes by just viewing two-dimensional imagery," Stokely said. "ArcGIS provides the ability to incorporate other geospatial data layers, such as contour lines and digital elevation models, to inform imagery interpretation. With orthomapping, the ability to view and pan around three-dimensional images on a computer screen is amazing. You know how crisp the imagery can be and then how the three-dimensional image can be so realistic."

Making decisions in 3D

In addition to producing more accurate images that are faster to review, aerial photography in a 3D format helps decision-makers see original imagery in a mapping context and identify relevant facts about authenticity.

Updated digital maps enhance but do not replace historical data. They create a new level of rigor and analysis. Compiling detailed information about a specific site for one or more time periods helps the EPA understand and document the site's current condition and guide decisions about its future. In addition, ArcGIS can help the EPA understand the impacts of climate change and human activities on natural landscapes. ArcGIS supports data visualization, analysis, and authoritative data maintenance.

"There are still things to discover in historical imagery about past pollution events where three-dimensional viewing is helpful," Stokely said.

A version of this story originally titled "3D Mapping Helps EPA Preserve Freshwater Resources" appeared in the Fall 2022 issue of *ArcUser*.

PART 4

CLIMATE SCIENCE

CLIMATE CHANGE IS A GLOBAL PROBLEM FELT ON LOCAL scales, threatening lives, infrastructure, and the economy. Rising seas, more frequent heat waves, extreme wildfires, degraded air quality, and significant floods and droughts are associated with the changing climate. As these events intensify, governments and stakeholders can benefit from a shared understanding of climate dynamics as they create safer, more resilient communities. Increasingly, researchers rely on GIS to understand the climate crisis by using tools to map climate characteristics, assess risk, plan mitigation projects and strategies, and coordinate response and recovery efforts. ArcGIS connects data, people, and predictive analysis with an interdisciplinary science-based approach to climate change. GIS helps stakeholders understand disparate climate data, develop information products, and activate resiliency plans. ArcGIS supports decision-making that can protect lives and property and build economic development.

The impacts of climate change are felt everywhere, from diminishing biodiversity to human, wildlife, and insect migration. But through individual and collective actions, humankind can create sustainable solutions. Using risk analytics and GIS technology, business and government leaders can respond to climate change using location intelligence and GIS.

Apply GIS to address the impacts of climate change

GIS technology is advancing climate risk modeling and resiliency planning. GIS facilitates geospatial data analytics and detailed 3D visualizations, bringing clarity to earth dynamics as governments, businesses, and organizations plan for and militate against climate threats.

- **Prepare for climate impacts:** Location intelligence helps build a detailed picture of potential climate hazards by combining climate science-based risk assessments with demographics, land cover, business activity, and societal data.
- **Perform and simplify analysis and discover patterns:** Climate professionals use GIS to investigate climate scenarios with 3D dynamic maps, time series simulations, and real-time interactive dashboards.
- **Turn data into actionable information:** Climate risk analysis, fueled by climate and weather data, AI algorithms, and location technology, connects predictions to places and assets to better understand climate impacts.
- **Restore, preserve, and protect:** Adopt a framework for iterative climate resiliency plans informed by predictive modeling and stakeholder engagement.
- **Raise awareness on climate risk:** Present the results of climate analysis, including possible solutions, to the public with interactive maps and multimedia for a visual narrative that connects with everyone.

GIS in action

Next, we'll look at some real-life stories of how organizations are using GIS to forecast, prepare for, and adapt to climate change.

ADDING PREDICTION TO PLANNING AS DISASTER COSTS RISE

Esri

THE COST OF NATURAL DISASTERS IN 2020 WAS NEARLY twice the previous year's total, increasing to $210 billion from $116 billion, according to a report from global reinsurer Munich Re. US losses totaled $95 billion, mostly from hurricanes, storms, and wildfires. Combined with NASA's analysis that 2020 tied with 2016 as the hottest years on record to date, the numbers point to accelerating climate risk.

The losses don't just affect insurers such as Munich Re; they also impact businesses and people. Most corporate directors already see climate risk affecting their organizations, according to Columbia Law School. The survey highlights motivations for directors to prioritize climate risk in business strategies—from managing cost and practicing corporate responsibility to identifying new business opportunities.

Companies are using technologies such as location intelligence and AI to predict potential climate risk and inform long-term planning. They're also moving toward real-time situational awareness to become more resilient when responding to natural disasters and other business disruptions.

Predicting climate-related risks

While no one can predict exactly where and when natural disasters will occur, companies are analyzing historical data to predict the most vulnerable areas. The analysis helps business leaders clarify long-term investments such as store locations, offices, manufacturing plants, and supply chain routes. For example, if location analysis shows that floods often affect an area or may do so in the future, supply chain managers can choose to ship or source goods another way, invest in resilient equipment and assets, or shift supply centers to safer areas.

Weather data, AI algorithms, and location technology fuel predictive climate risk analysis for places and assets.

Climate risk prediction is not an easy calculation. The accuracy and scale of data, precision of AI models, and expertise needed for such predictions shouldn't be underestimated. But partnerships in this task are becoming more common. When leaders at global telecom giant AT&T wanted to map the company's climate risk for the next 30 years, they partnered with Argonne National Laboratory; together they applied sophisticated AI models and GIS technology to produce results.

Monitoring disruptions for quick response

Beyond long-term climate risk management, business leaders apply location technology to improve short-term disaster response. The use of location intelligence often leads to business resilience when disaster strikes. As weather threatens, smart maps can provide companies advance notice of where threats exist, helping them adjust operations, measure impacts, and recover.

Imagine that a hurricane is moving toward a coastal area. A business informed by GIS-driven location intelligence would use a smart map to see how the predicted path might impact operations. Risk managers could understand, for example, which business locations fall within the affected area, the volume of sales at these locations, the supply chain partners that are at risk, where vehicles will be at the time of predicted impact, and how to reroute deliveries around or through the danger zone. After a disaster occurs, teams can swiftly measure the impacts and coordinate recovery efforts.

Businesses with data and technology in place have an advantage because they can more quickly adjust their operations. For example, risk management teams at global auto manufacturer General Motors integrated location intelligence into their operations so they can see how a disaster might impact the business—down to specific vehicle parts and their journeys across the global supply chain.

Planning a sustainable future

Continuing to build and rebuild in areas where natural disasters often occur is dangerous, especially with the frequency and cost of such events expected to increase with climate change. Enlisting predictive location intelligence in deciding where to build and invest isn't simply a matter of profit and loss; it can directly affect lives. Munich Re reported 8,200 deaths in 2020 from natural disasters alone.

The significant increase in climate impacts underscores the urgency of adopting changes quickly and at scale, supported by digital technologies such as AI and GIS. Corporate social responsibility to address climate change represents a worthwhile long-term strategy to mitigate the disasters that affect lives and balance sheets.

A version of this story by Alexander Martonik titled "As Disaster Costs Rise, Executives Add Prediction to Planning" originally appeared in *WhereNext* on February 2, 2021.

TAPPING ADVANCED ANALYTICS TO MAP DECADES OF CLIMATE RISK

AT&T

WHEN ONE OF THE WORLD'S BIGGEST COMPANIES USES big data and location intelligence to predict how climate change will affect its business for the next 30 years, the signal is clear: climate risk is real, and businesses must adapt.

Every year, the World Economic Forum asks leaders in government, business, and academia to identify the most consequential risks facing the world over the next decade. In 2020, for the first time in its 14-year history, the Global Risks Report said the top five risks were climate related, from extreme weather to collapsing ecosystems.

The economic fallout of climate change is apparent: real estate properties in danger of being underwater if sea levels rise one foot sold for 15 percent less than properties without flood risk. In a recent open letter to corporate executives, BlackRock CEO and chair Laurence Fink emphasized that climate risk would be a major guiding force in how the firm invests its more than $8 trillion in assets going forward.

In part, insurers and influential investors are driving the trend, demanding new levels of transparency from businesses on the climate risks their assets face. Industry leaders are paying attention because the costs of climate change become more apparent as its impacts accelerate. In 2017, when Hurricanes Maria and Harvey pummeled the United States and nearby island nations and storms battered Northern and Central Europe, insurers paid a record $135 billion.

Assessing and mapping climate risk

Leaders at AT&T—the world's largest telecom company by market cap—focus on mitigating climate damage even as they adapt their business to its impacts. A utility for millions of people, AT&T connects millions of people to home computers, mobile devices, Internet of Things (IoT) sensors, and more. Those services are digital, but the infrastructure that powers them—from cell towers to base stations—is vulnerable to climate impacts.

A public-private collaboration between the telecom giant and the US Department of Energy's Argonne National Laboratory in Lemont, Illinois, illustrates how data and location intelligence can identify climate risk and business adaptation to it. Drawing on the data-gathering power and supercomputing of the lab and the visual and analytic capabilities of GIS technology, AT&T developed a climate analysis tool that identifies areas of its network most at risk in the US Southeast.

With an unprecedented degree of detail, the utility forecast how a 50-year storm (which statistically has a 2 percent chance of happening in a given year) could affect infrastructure such as buildings and cell towers in Georgia, North Carolina, South Carolina, and Florida.

Bolstering climate resiliency through data

AT&T had been formalizing its climate adaptation efforts since 2015, but the cascade of severe weather that struck the United States between 2016 and 2018 showed the reality of climate change. During that time, repair and recovery from major storms cost AT&T, with its vast infrastructure network and thousands of pieces of equipment, more than $800 million.

In response, AT&T data and sustainability teams developed a tool that would predict impacts on areas of the network most at risk from climate change to the year 2050.

"One of the things we quickly realized was that if we're going to talk about climate change, probably the best thing we could do is give folks a visual representation of that," says Shannon Carroll, director of global environmental sustainability at AT&T. "You have to think about the end user. You could give them a bunch of datasets, but how useful is that really?"

To capture climate data in a visually engaging way, AT&T turned to the GIS technology it was already using to map the locations of its infrastructure assets. The smart map communicated with stakeholders and shaped decisions around adaptation and resilience.

Although the company relies on its own meteorological operation center to forecast short-term weather, it needed expertise from Argonne to build predictive models for climate change.

Groundbreaking new climate modeling

Companies that conduct climate research and adaptation analysis sometimes rely on generalized data that's often dated by several years or decades. By working with Argonne, AT&T had the advantage of accessing current data tailored to locations of interest and climate risk.

"AT&T's problem was very complex and very, very specific," says Thomas Wall, program lead for engineering and applied resilience at Argonne. "It was also at a systems-level scale that is much larger and more detailed than most of what I've seen for these types of projects."

The telecom giant focused on flood risk and high-intensity winds—the two threats deemed most significant to electrical and battery-powered equipment aboveground and underground. Company leaders wanted to know the depth of inundation in scenarios of varying severity in the four southeastern states of focus, inland and along the coast.

Argonne built on its physics-based regional climate modeling system, taking global climate predictions and applying them to a local level. While most climate modeling of this kind works on

12-kilometer blocks, Argonne focused on hyperlocal, 200-meter blocks—the most detailed level of climate modeling available in the four southeastern states.

Adaptation today for resiliency tomorrow

To address flood risks, researchers used advanced hydrologic and hydraulic models to simulate how water would flow and pool around terrain. The analysis was broken into 200-meter-by-200-meter cells—the size of seven football fields—and applied throughout the Southeast, covering more than 35 million grid cells. Using supercomputers, Argonne reduced the project data, equivalent to about 500 billion pages of text, to an output that could be formatted into AT&T's GIS. The utility then layered the climate forecasts and information on maps of the company's network assets.

"What's really important here is the quality of the data," Carroll says. "It's never perfect, but if we know with a 95 percent confidence interval that between now and the next 30 years, a specific grid cell will have a maximum flooding of 15 feet, that's really good information to have." Using the smart maps, AT&T can determine which assets they have in each cell and how susceptible they are to flooding.

The precision of the climate data and the visualization of smart maps enabled a sharper level of predictive analysis, helping AT&T build resiliency. Even within a mile or two, local topography, such as hills and valleys, affects coastal flooding and presents distinct levels of risk.

Understanding those risks, a planning team might move construction of a cell tower 200 meters south to an area less prone to floods or wind or to the other side of a highway. The team can strengthen existing facilities. One building may need sandbag reinforcement whereas another building may need to elevate its electric batteries above maximum flood levels.

Strengthening communities in the face of climate threats

From the beginning, the AT&T teams working on the climate risk analysis tool decided to make the data publicly accessible, publicizing the decision through press releases and social media channels that encouraged people and groups to download the data.

AT&T and Argonne invited local municipalities and universities in the four-state region to submit applications to use the data in ways that would address local problems. They selected five universities and gave each $50,000 to assist communities with climate resiliency and adaptation.

Part of Argonne's motivation in collaborating with companies such as AT&T on mapping climate risk is that the insights often benefit resiliency efforts in the larger region.

Businesses are increasingly stepping into the role of neighborhood watch, alerting communities to the impacts of climate change as part of their corporate responsibility. For instance, the design firm Atkins used GIS technology to develop a simulator that shows residents of a given municipality how the impacts of climate change will affect their daily lives.

The effort addresses the larger reality that no single company's efforts will address the challenges of climate change. But by adapting to risks, a company creates resources—data, processes, maps—that business leaders can share, encouraging orchestrated action in the most vulnerable places.

"I've got two young kids," said Antoine Diffloth, director of data insights and cybersecurity at AT&T. "It might sound trite, but I want to do what I can to leave a better place for them. I can't set policy, I'm not a climatologist. But I can do stuff with data. So, this is my contribution to leaving the world a better place."

A version of this story by Marianna Kantor and Jeffrey Peters titled "AT&T Taps Advanced Analytics to Map Decades of Climate Risk" originally appeared in *WhereNext* on June 30, 2020.

BUILDING A MARITIME SPATIAL ATLAS TO MANAGE CLIMATE CHANGE

Maritime and Port Authority of Singapore

THE FUTURE OF SINGAPORE'S COASTAL AREAS IS A MATTER of strategic global importance and local existential survival. The Port of Singapore, located where the Indian and Pacific Oceans meet, is the world's largest and busiest trans-shipment point. Much of the island rises a few meters above sea level, leaving it vulnerable to the effects of climate change.

To address the challenges of rising seas, more intense and frequent storms, and other threats, the Maritime and Port Authority (MPA) of Singapore manages GeoSpace-Sea, a national marine spatial data infrastructure of Singapore's coast and coastal waters. GeoSpace-Sea uses GIS to assemble and display layers of maritime and marine data that bring clarity to the cartographic frontier.

Singapore is the world's largest trans-shipment hub, where container connections are made to more than 600 ports across 120 countries.

Parry Oei, a hydrography advisor for MPA, is one of the prime movers behind the creation of GeoSpace-Sea to address change and prepare for the future:

- For scientists, GeoSpace-Sea is a way to understand how the world's oceans are changing and how the changes are impacting Singapore.
- For planners, GeoSpace-Sea is a way to anticipate the changes and make predictive analysis using a shared repository of information.
- For port officials, GeoSpace-Sea provides higher-resolution data to underpin operations, allowing them to see and consider the consequences of decisions such as the impact of dredging on sea life.

Everything is connected

Oei traces GeoSpace-Sea's provenance to the late 1980s, when he was working in Hobart, Australia, as part of the Asia-Australia Project, a collaborative effort to understand the emerging signs of climate change.

It was during this time in Australia that Oei began to think about mapping the complexity of the ocean. Climate change would surely impact the oceans, and the oceans would in turn affect climate change.

"The sea is a planet that has not yet been fully explored," Oei said. "It's quite mind-boggling how everything is linked to everything else."

The questions were how to gather data and what to do with it. At that time, GIS was a niche technology used primarily by trained GIS experts. Because this group did not include hydrographers, Oei found little enthusiasm for this use of GIS within Singapore's government.

The project takes shape

As GIS became a cloud-based technology, it drew cross-discipline collaborations and began to include spatial data at all scales without limits on data volumes. By the mid-2000s, as climate change became more of a pressing concern, and sea level rise became an existential threat to Singapore, Oei's idea to map the ocean gained momentum.

"For the longest time, we had been trying to make the case for using GIS for marine protection and to study the dynamic nature of the seabed, but the question was how to do it," Oei said. "Climate change gave us the profile and attention we needed."

The challenge of extreme weather

Predictive analysis indicates that climate change will bring more intense storms, with storm surge and inland flooding.

As a marine geospatial knowledge repository, GeoSpace-Sea can increase understanding of extreme weather. For instance, until recently it was assumed that water flow in the Singapore Strait was

Viewing the coastline in 2D within GeoSpace-Sea gives a detailed view and interactive tools for markup and measurement on the map. Screenshot courtesy of the Maritime and Port Authority of Singapore.

mainly affected by the Northwest Monsoon, the period in winter and early spring when the Indian Ocean's trapped heat draws cold air from the Himalayas. Scientists now believe the Southwest Monsoon—the summer season when winds blow moist air from the ocean toward the Himalayas—wields a stronger influence.

"The question will be 'How do we get rid of water when it floods, and how do we prevent seawater from coming in?'" Oei said. "I have no easy answers, but I do believe we need to monitor everything in the environment—weather patterns, temperature, sea level rise—because it's all part of climate change."

The future of GeoSpace-Sea

Oei hopes to add more weather data to further this integrated approach. The MPA, which opened the data portal to the public in 2022. The portal also serves as an unfolding historical document that lets researchers study past trends to predict future developments. Some datasets date more than 20 years, and new data is added as it becomes available.

The portal planned to increase the use of sensors and the IoT and have some of the data update in near real time. "With modeling, there's only so much you can do," Oei said. "But if you integrate modern sensors, this truly becomes a digital twin and you become more proactive than reactive."

GeoSpace-Sea also plans an AI aspect, using machine learning protocols to search for patterns and hot spots. As an example, Oei suggests that GeoSpace-Sea could help Singapore better understand the conditions that lead to red tide, when algal blooms wreak havoc on its coastal ecosystems.

"What makes AI so interesting is that it could uncover things we don't know, but we honestly can't say what it will show us," Oei said.

The ultimate goal of GeoSpace-Sea, he explained, is to further scientific progress by gathering as much information as possible to address questions we may not have thought of yet. Understanding something as complex as how Singapore's connection to the ocean is evolving requires technological sophistication and constant collaboration. "Alone, we can go fast," Oei said. "But together we can go far."

A version of this story by Chris Fowler titled "Singapore Builds a Massive Maritime Spatial Atlas to Manage Climate Change" originally appeared in the *Esri Blog* on July 7, 2022.

CLIMATE CHANGE PROMPTS GRENADA TO CREATE NATIONAL DIGITAL TWIN

Esri

GRENADA, A NATION SMALL IN SIZE AND POPULATION, became one of the first countries to make a digital copy of itself in 2021, creating a 3D model that government officials can use for sustainability plans.

Like many island nations, Grenada confronts an uncertain future in the face of climate change. Increasing heat, intense rainfall, and saltwater intrusion into the water supply and soil have begun to threaten the country's two primary economies—agriculture and tourism. One challenge was how to continue to grow in a sustainable way and adapt to the changing environment. Addressing this challenge required a geographic approach—understanding what was happening and where.

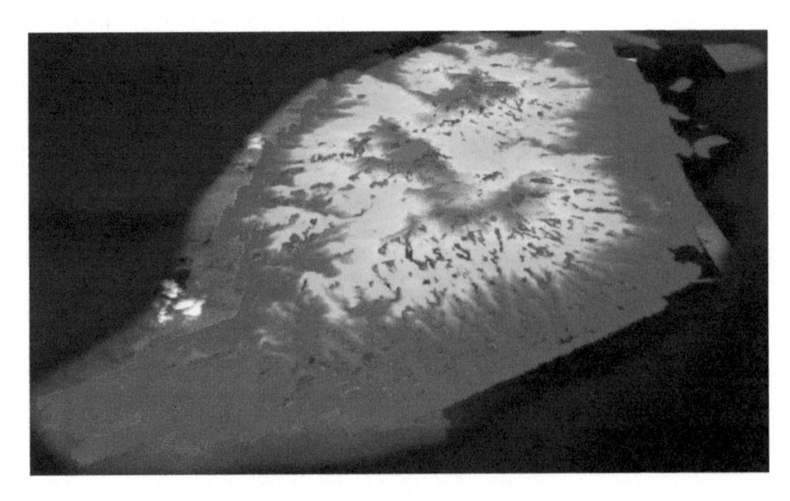

Airborne lidar over Grenada, colored by point-cloud elevation.

Grenada's government had stores of raw geospatial data in the office of the Ministry of Agriculture and Lands. In 2019, the office received World Bank funds through the Regional Disaster Vulnerability Reduction Project and hired Fugro, a company that specializes in geographic and geologic data gathering and analysis, to do extensive aerial reconnaissance of Grenada. Fugro surveyed the Caribbean nation's three major islands, as well as six smaller ones. The result was a trove of information, including a lidar point cloud and extensive aerial images. But for practical purposes, there appeared to be no way to organize all this valuable information until the ministry used GIS technology to create a digital twin.

The power of two

The digital twin—a virtual representation of the objects and processes of a real-world system—has rapidly evolved in recent years. The earliest digital twins were built to monitor the functioning of industrial factories, even to the level of individual valves and gaskets. Digital twins are now complex enough to model entire municipalities. City managers use them to monitor urban functions. Planners use them to visualize and analyze the effects of proposed changes.

Today's digital twins can be intricately detailed. Singapore's digital twin, for instance, extends to underground infrastructure and even includes some indoor features.

GIS experts in Grenada's government decided to extend the country's twin nationwide. Necessary for making digital twins fully operational, a GIS stores and displays disparate datasets that share locational components. These interactive and collaborative 3D models can then be used to drive better decision-making and policies at a larger scale than previous systems allowed.

A GIS enabled officials in Grenada to stack the imagery and point cloud data. The imagery and data could be consumed as separate map layers and combined to create something functionally larger than the sum of its parts. With its 20-centimeter resolution, the resultant aerial imagery produced a detailed representation of the island. Linking the 3D lidar data brought the imagery into full relief.

The digital twin goes deep

Government officials sought to use the country's digital twin to improve the lives of its residents, who are at the mercy of a swiftly changing ecosystem. The work started with extracting streets and buildings from the visual data so they could be sorted and quantified. The data could then be manually coded, but that process could require up to six months of work.

Grenadian officials worked with analysts from Esri to deploy AI capabilities within a GIS. A deep learning model was used to identify buildings. Within a day, analysts used the program to extract and label 55,000 built structures. They used geospatial artificial intelligence (GeoAI) capabilities to sort and classify other parts of the digital twin's visual data, such as roads, powerlines, streams, and other inland bodies of water, along with vegetation and land cover.

These classifications are valuable by themselves. For instance, staff from Grenada's Central Statistics Office, which partnered in the digital twin efforts, realized that the building data could simplify the process of planning the national census and recognized the value of having a complete building inventory of the country for the first time.

Combining the data categories created synergy. The Grenadian government and Esri used stream data, vegetation classifications, and digital terrain modeling (another segment of Fugro's aerial collections) to highlight spots in the country most in danger from landslides. This process was mainly automated: with classifications in place, the GIS generated the results. Other formulas and calculations produced flood susceptibility models, revealing where island

Projection of two-meter sea level rise in St. George Harbor, Grenada.

residents were most vulnerable to extreme weather brought on by climate change.

The 3D nature of the lidar data contributed to the use of Grenada's new model. Seeing how far a building or road is from a landslide-prone area is helpful. Having the ability to zoom in and examine how a building perches on a steep hillside or how a vulnerable road's angle of descent would appear from the perspective of a motorist, pedestrian, or cyclist adds context.

Seeing the future

Grenada's digital twin has a foundational quality. The data it comprises is now the basis for what the UN calls an integrated geospatial information framework (IGIF). It provides a complete view—realistic and integrated—of the country, which supports decision-making. This digital twin also has a predictive component that allows officials to visualize future challenges posed by climate change, along with solutions.

A dashboard tracks enumerator progress for a population and housing census.

The government has used the digital twin and bathymetry information from Fugro to model scenarios for sea level rise—including storm surge and flooding damage—to see what will be impacted and where. The visual context of the map transcends numeric projections, facilitating policy making for prevention and mitigation.

The digital twin can also serve as an ongoing historical record. For example, the lidar data identified 4.5 million trees. If more aerial data is gathered at points in the future, the twin's GIS can analyze tree growth and note any significant deforestation. As important as the AI capabilities are for this kind of calculation, they wouldn't have been possible without the lidar-enhanced imagery. Grenadian planners, interested in growing the country sustainably, can now look at a section of the map and imagine how further development will impact—and be impacted by—future changes in vegetation.

The value of a GIS-powered digital twin is that it enhances human observation. Although a digital twin can't literally see into the future, it is a window into several potential futures. However, none of this would be possible without accurate imagery data and geospatial technology that ties geographic information together. These collaborative technologies are turning data into something meaningful, viewable, and measurable.

Working toward sustainability goals of their own, other countries will likely follow Grenada's lead, building location-intelligent digital twins of their own.

A version of this story titled "Climate Change Prompts Grenada to Create First National Digital Twin" by Linda Peters originally appeared in the *Esri Blog* on June 16, 2022.

MAPPING CLIMATE RISKS HELPS FACILITATE ACTION

New Mexico's Interagency Climate Change Task Force

LONG BEFORE SPANISH AND MEXICAN SETTLERS ARRIVED in what is now New Mexico, people had farmed the land for centuries. They devised sustainable dry-land farming practices that included irrigation canals and fields with porous volcanic pumice to absorb water and slowly release it.

Today, drought and rising temperatures are causing wells and drainage ditches to go dry on that land, forcing tough decisions about how to adapt. In some areas of the state, early shutoff of irrigation water and the rising cost of fuel and fertilizer has left some farmers in peril of losing their land.

To help New Mexicans prepare for uncertain climate outcomes, a team of scientists, technologists, and resource specialists launched the New Mexico *Climate Risk Map*. Experts at Energy, Minerals and Natural Resources Department (EMNRD) teamed with the Earth Data Analysis Center (EDAC) at the University of New Mexico. Using GIS, they aggregated and analyzed data to create the interactive map.

The map allows anyone to explore the worst outcomes of rising temperatures—drought, heat, flooding, degrading air quality, and wildfire—noting where and who will face heightened risks from these hardships. Communities can use the map to visualize climate pressures and see where they must decide to adapt.

Taking climate action against compounding problems

In addition to drought issues of 2021–22, the average annual temperature has increased three degrees since 1970 and is projected to

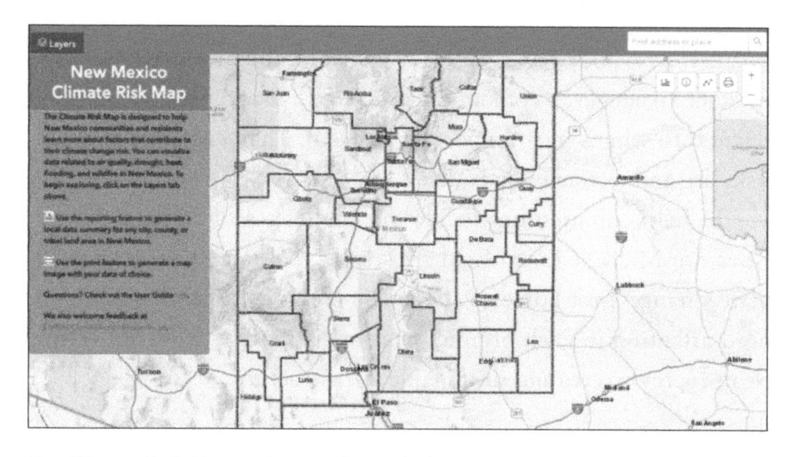

The *Climate Risk Map* is designed to help New Mexico communities and residents learn more about factors that contribute to their climate change risk.

rise another nine degrees by 2100. Together these changing conditions have increased wildfires.

On May 2, 2022, thousands of residents were evacuated from the city of Las Vegas in northern New Mexico as high winds blew embers ahead of the Calf Canyon Fire. The Calf Canyon Fire had merged with the Hermits Peak Fire, reaching megafire status as the largest fire in state history.

"Fires can harm habitat and create conditions that cause major floods that impact drinking water quality," said Maria Lohmann, who coordinates the New Mexico Climate Change Task Force. "It's really important that we start to talk about these repercussions, because New Mexico has some really special and unique landscapes that are already struggling."

In addition to the loss of habitat, fire damage poses an immediate threat when rainstorms carry away soil no longer held back by vegetation. This debris can clog channels and culverts, fill reservoirs and retention ponds, and destroy fields and crops.

"The strength of this map is its ability to run different scenarios, because everybody has a different focus," Lohmann said. "Users—whether an agency, a community group, or a local government—can relate it to what's most important to them. They can see where the risks are and examine the issues."

To help communities tackle the challenges of climate change, Governor Michelle Lujan Grisham created the state's Interagency Climate Change Task Force in 2019. The group focused on issues that need attention in each of nine focus areas ranging from delivering clean energy to creating sustainable infrastructure.

More heat, less rain, and the need for cooling

"With extreme heat coming, energy has to be dependable and affordable so people can cool themselves," said Robert Gomez, resilience coordinator for EMNRD's Sustainability and Resilience program. "Those kinds of connected factors have to be built into our ongoing resilience and adaptation plans."

The changes in climate and high fuel loads create the conditions for catastrophic fires. "We have really set ourselves up for some potentially serious situations there," Gomez said.

In New Mexico, and across the greater Southwest, fuel conditions, along with projected increases in seasonal temperatures and decreases in annual precipitation, increase the potential for more frequent, intense, and extended wildfires.

Creating the interactive map

The *Climate Risk Map* helps users visualize impacts statewide. People can choose from a list of different climate hazards and sensitivity factors to view those layers on the map. It can also generate specific reports of interest by combining details; calculating statistics; generating tables, charts, and maps; and then producing a PDF file that can be saved and shared.

The map's climate risk tool emphasizes equity and addressing the needs of overburdened communities, including demographic factors such as race and poverty levels. So far, the task force has used it to examine equity and create five-year action plans. The map is also available for community planners, tribal nations and pueblos, and anyone interested in understanding their fire risk.

The map can also support other agencies such as New Mexico's Department of Health, which responds to climate-related health effects including respiratory and other conditions due to wildfire smoke and for the prevention of heat-related illness. With the tool, they can also examine equity. "Health outcomes of climate change are important questions we need to ask ourselves," Gomez said. "And specifically examining who in which areas are particularly vulnerable."

A version of this story by Patricia Cummens titled "New Mexico Maps Climate Risks to Inform Climate Action" originally appeared in the *Esri Blog* on May 17, 2022.

ADDRESSING CLIMATE CHANGE AS A COMPANY-WIDE PRIORITY

Esri

CLIMATE CHANGE HAS LED TO MORE FREQUENT AND extreme weather, along with gradual, longer-term changes to the environment, and those changes are impacting people's lives.

According to the World Meteorological Organization, 2021 was one of the seven warmest years on record—and the last seven years have been the warmest in history. This was punctuated with record drought in South America and unprecedented flooding in Asia and Europe. Some of the most extraordinary weather in 2021 didn't occur in typically warm areas. Canada shattered its record-high temperature by hitting 49.6 degrees Celsius (121 degrees Fahrenheit) in British Columbia in June 2021.

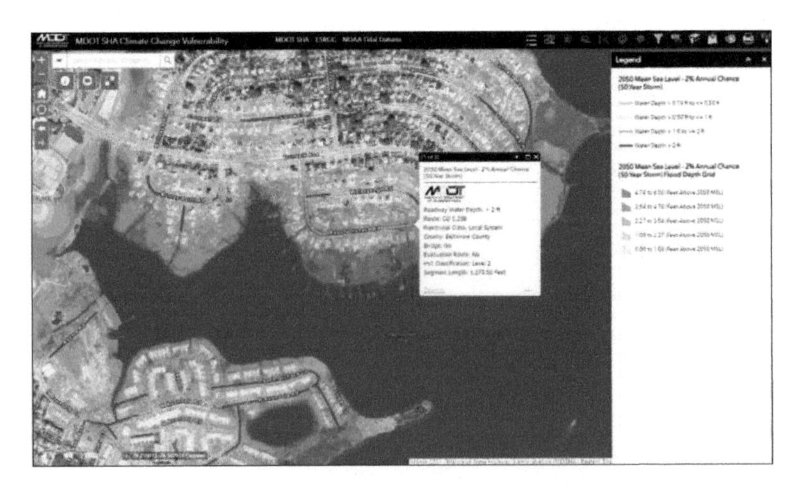

Having access to various climate scenarios—such as the potential extents of sea level rise during a 50-year storm, seen here for parts of Maryland—can help planners and designers prepare for future climate conditions.

No place in the world is immune to the effects of climate change. Combating this crisis requires new ways of thinking and behaving. To understand changing climate conditions, anticipate future impacts, and drive mitigation and adaption strategies across all aspects of society, people need access to sound data, tools, apps, and workflows. This work will require unparalleled collaboration—and a geographic approach.

Champions of climate change are already taking action

In early 2021, one of US President Biden's first acts in office was to sign a comprehensive executive order on climate change that calls for a holistic, science-driven approach to confront the crisis. At the end of 2021, global leaders, scientists, and young people attended the UN Climate Change Conference, COP26, to commit to collective action in response to climate change. The European Green Deal, which seeks to make the European Union (EU) climate neutral by 2050, strives to engage all people in climate action.

Governors, mayors, and community leaders worldwide are developing climate resilience plans:

- Local governments are evaluating new infrastructure projects for future climate risks, retrofitting buildings to improve energy efficiency, and taking a geographic approach to siting areas for wind and solar energy generation. They are also mapping where to equitably install EV charging stations and establish interstate EV corridors.
- Private companies are evaluating climate-induced vulnerabilities in their facilities and supply chains to protect their businesses and be more responsible corporate citizens.
- Delivery companies are analyzing driving patterns and location characteristics to evaluate how to electrify their fleets.

- Architecture and engineering firms are applying climate-smart principles to their project designs.
- Health-care providers are rethinking public health issues as climate change alters the way diseases spread.
- Nonprofit organizations are working to conserve land and water, enhance nature-based climate solutions, and help governments and businesses harness ideas and technologies.

New climate initiatives at Esri focus on collaboration

For more than 30 years, Esri has collaborated with users in the climate community, ranging from climate researchers to staff at national weather agencies. In the early 1990s, for example, Esri teamed with the National Science Foundation in the United States to study the hole in the ozone layer and model Antarctic solar radiation. Esri's work on climate-related initiatives has expanded over the years to the point that the company was awarded a top ranking in the Forrester New Wave: Climate Risk Analytics, Q3 2020 report.

Recognizing the growing urgency to scale up its climate efforts, Esri formed a new climate team, pulling together experts from across the organization to institute a comprehensive approach to tackling the crisis. Out of this came five focus areas that are reshaping Esri's work:

- Mitigate climate change.
- Understand the impacts of future climate conditions.
- Assess risks to people and property.
- Geodesign with climate awareness.
- Communicate to educate and spur action.

Mitigate climate change

Esri's climate initiative begins with climate mitigation measures that support efforts to reduce greenhouse gas emissions. These measures build on Esri's previous work to help find sites for alternative energy production, model transportation alternatives, and assist with sustainable land and forest management.

Carbon capture company Summit Carbon Solutions, for example, recently adopted ArcGIS technology to help with its efforts to trap and store carbon dioxide before it is emitted into the atmosphere. In addition, the US Department of Agriculture and the EU use ArcGIS to map areas that have the highest potential for carbon sequestration to determine investments in farm management techniques.

Esri has installed solar panels at its campus in Redlands, California, to reduce its carbon footprint.

Understand the impacts of future climate conditions

To prepare for future climate conditions, planners and designers must access future climate scenarios and impact modeling results. Twenty years ago, Esri collaborated with a team at the National Center for Atmospheric Research (NCAR) to incorporate netCDF (network Common Data Form) into ArcGIS and help make it an Open Geospatial Consortium Inc. (OGC) standard.

Esri will continue to make climate projection model outputs easy to use and is committed to helping others understand how to use this data. Layers slated to be improved and updated include projections of future temperature, precipitation, sea level rise, and inland flooding extents and frequency—all of which can be used to model future climate-related impacts. Esri will also expand related tools and workflows.

Assess risk to people and property

Impact maps that show where flooding and other weather-related events may occur in 2050 can serve as effective decision-making tools when they are overlaid with population projections, planned infrastructure, ecosystems, and other data. For example, analyzing national heat indexes can help decision-makers see where to place cooling centers and plant trees.

Esri is developing new workflows, ArcGIS Hub℠ solutions, and lessons to streamline risk assessments and help governments and businesses prioritize critical areas for intervention. The Esri partner community also provides climate-focused solutions and services, such as the climate risk data products from Mayday.ai and tailored climate resilience planning from FernLeaf Interactive.

Geodesign with climate awareness

Designing and developing spaces is always an exercise in balancing competing goals. Assessing current and future climate risks is part of that consideration.

To help planners, resilience officers, and those who are not GIS or climate experts make well-informed decisions, Esri is developing replicable solutions and publishing data that focuses specifically on climate issues. These solutions are rolled out alongside training programs that show users how to use content from ArcGIS Living Atlas of the World along with their own high-resolution data.

The strength of this approach is in partnering with experts to organize relevant climate data for use through apps and workflows. For instance, for the Climate-Smart Communities challenge sponsored by The Opportunity Project, Esri created a Hub template for extreme heat with data and tools that guide users through five steps for making resilience plans. Esri will use this resource pattern for other threat themes, including wildfires and floods.

Communicate to educate and spur action

The GIS community needs to engage with the public to prompt action regarding the challenges for progress that lie ahead.

In the United Kingdom, Esri UK teamed with the country's Meteorological Office to create GIS-based lessons for K–12 students that help them interpret complex scientific climate data. The idea is to show young people how they can learn about climate change and motivate their peers to act to address it.

Another tool for communicating ideas is ArcGIS StoryMaps℠, a story-authoring app that helps users build narratives around maps and data. Esri also works with Project Drawdown, a nonprofit dedicated to reducing greenhouse gas emissions, to add interactive mapping experiences and apps to its two educational series, Climate

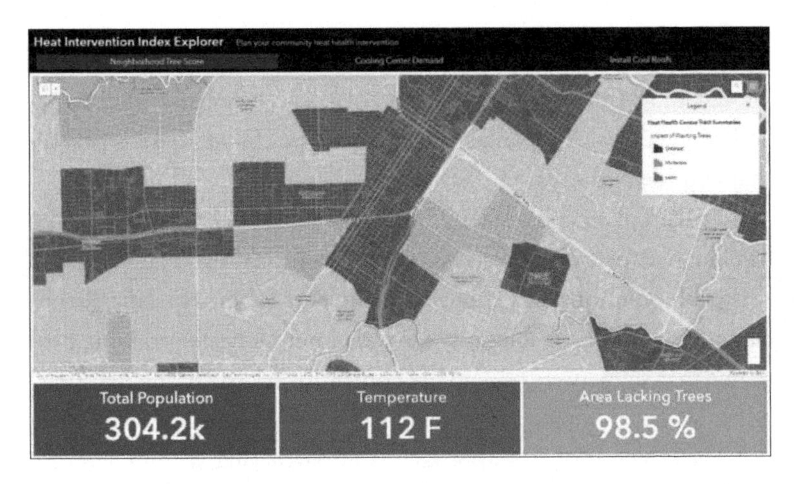

Analyzing national heat indexes can help local government decision-makers see where to plant trees.

Solutions 101 and Climate Solutions 201. The online courses helps leaders in business, philanthropy, investment houses, and communities think strategically about climate solutions.

Technology is a crucial part of the process

Esri's biggest contribution to fighting climate change is supporting GIS users in efforts to address climate change. Esri collaborates with users across sectors to learn about their challenges, workflows, and new opportunities to act. "We recognize the seriousness of the climate crisis," said Jack Dangermond, president of Esri, "and we know full well that technology will be a crucial part of the solution."

A version of this story by Patricia Cummens and Steve Kopp titled "Esri Addresses Climate Change as a Top, Company-Wide Priority" originally appeared in the Spring 2022 issue of *ArcNews*.

THE BUSINESS EXECUTIVE'S BLUEPRINT FOR SUSTAINABILITY

Esri

THERE'S A NEW KIND OF ENVIRONMENTAL WATCHDOG OUT there.

A decade ago, their predecessors walked through jungles and woods to document deforestation from illegal mining, oil drilling, cattle ranching, and palm oil farming. What once took months or years to uncover, today's watchdogs reveal in a day, using satellite imagery and AI-powered GIS technology.

Matt Finer, former senior research specialist at the nonprofit Amazon Conservation, helped bring these techniques into mainstream use. Today, the organization is infusing the role of global watchdog with more precision and speed.

A watchdog peers deep into the supply chain for Amazon Conservation

For global businesses, total supply chain visibility is rare—studies say just 6 percent of companies have achieved it. But executives can no longer plead ignorance of the activities occurring deep in their supply networks. New and pending legislation cracking down on deforestation, for instance, shows that the actions of any player in the supply chain—however minor or far from headquarters—are now a corporate risk.

Matt Finer's name is well-known among environmental monitors. In 2015, he founded an Amazon Conservation project to fine-tune and fast-track the monitoring of conservation land. Instead of relying on field crews, Finer and his Monitoring of the Amazon Project (MAAP) team use satellite imagery and GIS analysis to spot illegal deforestation. In the early days of the project, they created yearly reports based on 1,000-meter-resolution imagery. Now, they spot encroachments in areas smaller than half a tennis court—within a day or two of their occurrence.

Finer receives up to a thousand alerts daily indicating that a given pixel in satellite imagery has changed color since the flyover a day earlier. The change could indicate that a few trees may have been felled or that smoke may be rising. GIS software saves the MAAP team from manually investigating each alert. GIS groups pixels into clusters of activity, guiding MAAP and Amazon Conservation researchers to the most urgent areas of deforestation.

This watchdog work is happening at a time when the Amazon is in danger of losing its status as the planet's carbon sink. The World Wildlife Fund (WWF) describes the Amazon as a "vast biome that spans eight rapidly developing countries—Brazil, Bolivia, Peru, Ecuador, Colombia, Venezuela, Guyana, and Suriname—and French Guiana, an overseas territory of France." The landscape covers about 20 percent of the world's liquid freshwater and 2.7 million square miles in the Amazon basin, about 40 percent of South America, according to WWF.

Scientists say tree loss and fires threaten to turn the Amazon into a net emitter of carbon dioxide, which would hasten atmospheric warming and worsen climate change.

Finer and his team watch for activities that upset the natural balance. With the evidence they've compiled, authorities throughout the Amazon have shut down activities ranging from illegal palm oil production to destructive cattle ranching in endangered areas to illicit gold mining.

New maps empower the protectors

Nadia Mamani was born and raised in the biodiverse Madre de Dios region of Peru, where gold miners and others have illegally cleared the rainforest to set up operations.

Despite having lived in Madre de Dios all her life, she hadn't grasped the extent of the illegal mining until she began using GIS and remote sensing tools to complement her field investigations. What she saw inspired her to work for conservation and restoration of landscapes in Indigenous communities.

That newfound awareness parallels what a business executive might see when GIS-based insight reveals unwanted activity deep in the supply chain, otherwise out of sight.

"In my hometown, there are some areas that are not very accessible," Mamani said. "So for me, GIS and remote sensing gave me new eyes—a view from a different angle."

During her graduate studies in GIS and conservation, Mamani relied on Finer's maps to investigate mining activity in her town. Upon moving to the US to start her career, she started work as a GIS and remote sensing specialist for MAAP.

"After I moved here, I decided to work for my country and my hometown. And in some way, that's what we do—protecting my people, our forest. It's my home, and that's something that unifies me and Matt."

A pioneer in the right place at the right time

Several years ago, Finer and the MAAP program uncovered a massive deforestation project—2,000 hectares cleared in a couple of months by a cacao producer.

The team used GIS software to combine satellite imagery with data on the location of roads, country boundaries, mining concessions, and protected land.

The team showed its effectiveness when *El Comercio*, Peru's oldest and most popular newspaper, printed the satellite images of the cacao operation on its front page. When authorities broke up the operation, *El Comercio* printed that news, too.

Cat and mouse in the jungle

While watchdog groups such as Amazon Conservation were using technology to detect and expose threats, others were going undetected.

Sidney Novoa, director of GIS and technology at Conservación Amazónica, Amazon Conservation's partner organization in Peru, watched the process unfold. His early work focused on illegal timber operators that built huge roads and extracted vast swaths of wood from the rain forest. Smaller operations that worked as discreetly as possible, using small trucks and removing only select trees, often escaped notice.

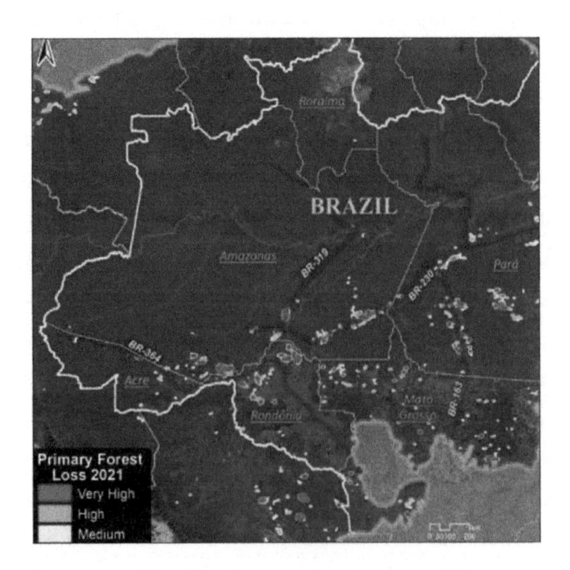

A GIS-based map produced by Amazon Conservation shows areas of concern.

Fortunately, the ever-increasing clarity of satellite images and a GIS technique called GeoAI gave the MAAP team a new advantage. When loggers try to thin trees in small numbers, the activity creates subtle but telltale gaps in the canopy.

That characteristic can be identified by machine learning algorithms once the GIS program is trained on the shape of the gap and, sometimes, the color of the area that the operation leaves, Novoa said. In other words, GeoAI reveals what some people hope will remain hidden.

"We identified camps, we identified the timber wood ready to be transported up to other areas," said Novoa. "We reported that to the local authorities here in Peru."

As pressure on the business world grows, more corporate executives will rely on watchdogs such as Finer—or location analysts within their own organizations—to root out bad practices from supply chains.

Although Finer and his MAAP colleagues address threats across the Amazon, Peruvian gold mining helps illustrate the team's impact and value of environmental monitors to change business practices.

"We've really been able to track that problem in real time, send that information to the Peruvian government, and really help the Peruvian government clamp down," Finer explains. "And we've seen a major decrease in illegal gold mining deforestation in the Peruvian Amazon. So it's really helping propel this field of real-time monitoring."

A version of this story by Alexander Martonik titled "GeoAI, Corporate Responsibility, and the Vigilance of a Climate Watchdog" originally appeared in *WhereNext* on May 17, 2022.

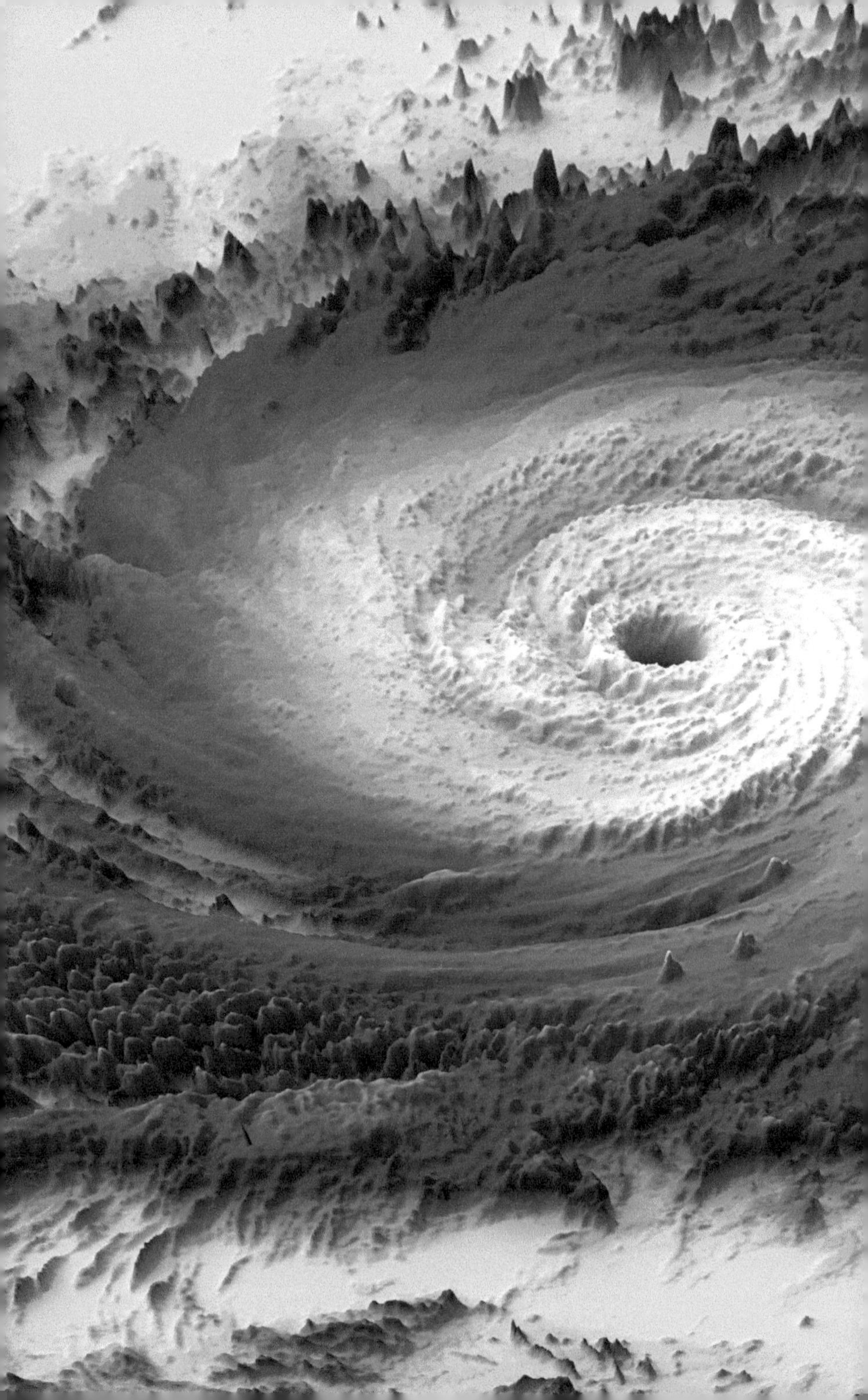

PART 5

WEATHER

TODAY'S EXTREME WEATHER INCLUDES AN INCREASE IN the number, duration, and intensity of storms, floods, heatwaves, droughts, and wildfires. In the United States alone, 2021's extreme weather cost $145 billion, with a loss of nearly 700 lives, according to the National Oceanic and Atmospheric Administration (NOAA).

Meteorological observations and research needed to produce the weather, climate, and water-related forecasts, predictions, and information products are becoming interdisciplinary, using data from demographics, social sciences, and environmental applications that standard weather analysis packages cannot fully manage.

As weather becomes more volatile and costly, society, businesses, and governments rely on atmospheric science and GIS products to support decision-making. GIS integrates meteorology with other physical, social, policy, and economic contexts to advance weather data visualization, prediction, and real-time decision-making. ArcGIS enables the meteorological community to blend troves of weather data with curated, location-based information. By upgrading an organization's geospatial resources, decision-makers can protect lives, property, and economic development.

New insights into weather services

ArcGIS connects data, processes, and people for an inclusive approach to understanding our changing weather. Next-generation GIS weather tools improve weather forecasting, climate monitoring, and reporting. More capable tools, workflows, and data access provide insights into weather patterns, forecasts, and predictions faster and in higher quality. More people can access, visualize, and analyze weather data through cloud-based web tools and data access services that ArcGIS offers.

Creating a common operating picture

ArcGIS provides a common geospatial framework to connect data and information of different spatial resolutions in one place. Visualization tools help users explore observations and reveal previously unseen patterns. ArcGIS allows meteorologists to analyze multidimensional weather and climate data and present more informative forecast products than the often-traditional static images that weather forecasters produce. With GIS, forecasters can aggregate vast amounts of data from time series collections to live feeds and visualize weather scenarios in dashboards and on-the-fly animations. GIS improves weather hazard warnings and watches and enhances communication between forecasters and the public through an open, extensible, and built-in collaborative structure:

- **Access weather data collections:** Securely provide access to streaming curated atmospheric datasets with ArcGIS. Automatically update weather maps and databases and respond with real-time alerts.

- **See weather patterns more clearly:** Use GIS to better understand weather scenarios with 3D dynamic maps, time series simulations, and real-time interactive dashboards that enhance visualization and uncover patterns.

- **Bring clarity to troves of data:** Integrate out-of-the-box modeling environments with open-source spatial algorithms, data libraries, and programming languages. Enrich research with ready-to-analyze and streaming weather data sources.
- **Make decisions confidently:** Discover real-time weather data, make predictions, and interpolate surfaces. Use streaming real-time weather data for more accurate weather predictions.
- **Operate securely with other scientists:** Access weather information across systems and share results and research with colleagues using a built-in collaborative framework.
- **Get stakeholders involved and excited:** Use a framework for iterative weather planning scenarios informed by predictive modeling and endorsed through stakeholder engagement.

GIS in action

Next, we'll look at some real-life stories of how organizations facilitate weather-driven decisions with GIS.

A GIS-DERIVED CLIMATOLOGY OF HAIL

Esri

HAIL IS FROZEN PRECIPITATION THAT FALLS IN THE FORM OF balls or irregular lumps of ice. It damages crops, livestock, trees, utility lines, and property and can cost billions of dollars in damage in a given year. In Minnesota, a hail storm in June 2017 caused about $2.5 billion in damage. Understanding where and when hailstorms (or events) occur can help local emergency managers, insurance assessors, and agricultural agents estimate the costs and mitigate the risks of hail damage.

This report explores the analysis of spatial and temporal patterns of hail for the contiguous US for 21 years from 1996 through 2016, thus generating a climatology of hail. A climatology is a summary of weather over a period of time. Data was gathered from the NOAA National Climatic Data Center's (NCDC) Storm Events Database. The database records the occurrence of storms and other significant weather phenomena having sufficient intensity to cause loss of life, injuries, significant property damage, and disruption to commerce. This report describes a study of severe storm climatology rather than a comprehensive meteorological climatology.

What causes hail?

Hail is associated with thunderstorms in which air is rapidly forced upward (updrafts). These updrafts push water vapor and rain high into the thunderstorm where the temperature drops below freezing, allowing ice to form around dust or other particles in the air. These particles serve as the seeds for hailstones, which fall to the ground once they become too heavy for the updrafts to keep them aloft.

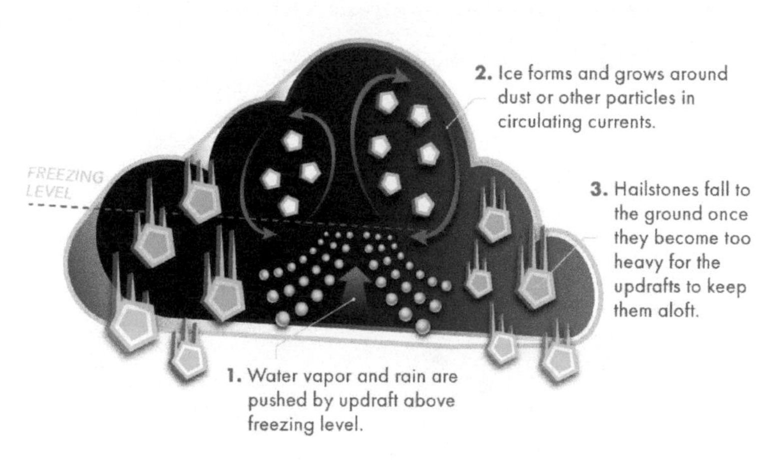

The process of hail formation.

A spatial perspective

The contiguous United States had more than 272,000 hailstorms from 1996 through 2016. Mapping every event produces an uninformative solid mass of points over the entire country. However, calculating a kernel density produces a continuous surface that shows the number of hailstorms per five-kilometer cell at any location.

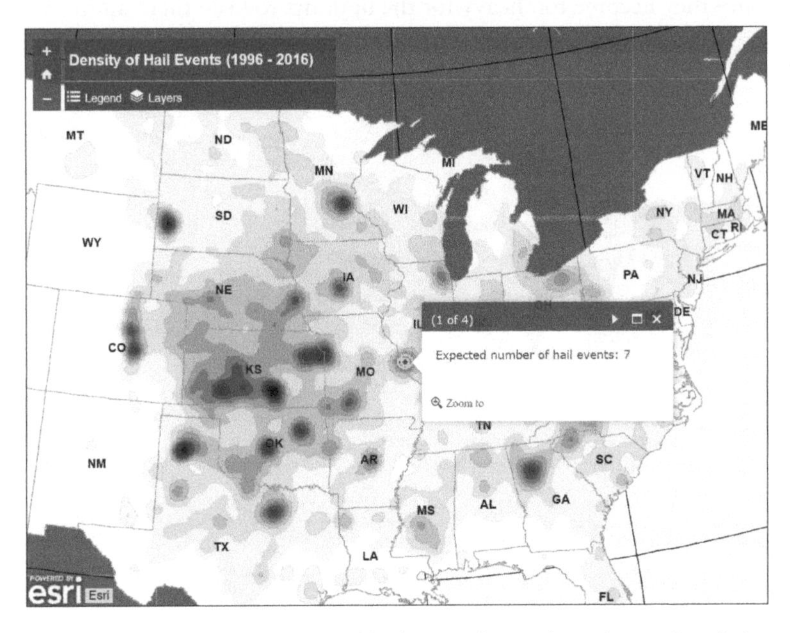

Results of kernel density analysis of hail events (1996 through 2016), with the darker purple showing greater intensity.

The map shows high numbers of hailstorms in the Great Plains. These storms are shown cumulatively, summing the events across time within each five-kilometer cell. This map helps answer the question, "What is the overall cumulative frequency of hailstorms in the United States?" The map indicates spatial patterns and allows for more detailed analyses.

A temporal perspective

A climatology summarizes weather variability over time, requiring a temporal perspective—in this instance, through visual analytics using a data clock. A data clock summarizes temporal data in two dimensions to reveal seasonal or cyclical patterns and trends. The data clock in the graphic showing severe hailstorms summarized over years confirms patterns with little hail occurring in January and February, increasing in March and April, the highest occurrence in May and June, moderate occurrence in July and August, and finally little activity in the fall and winter.

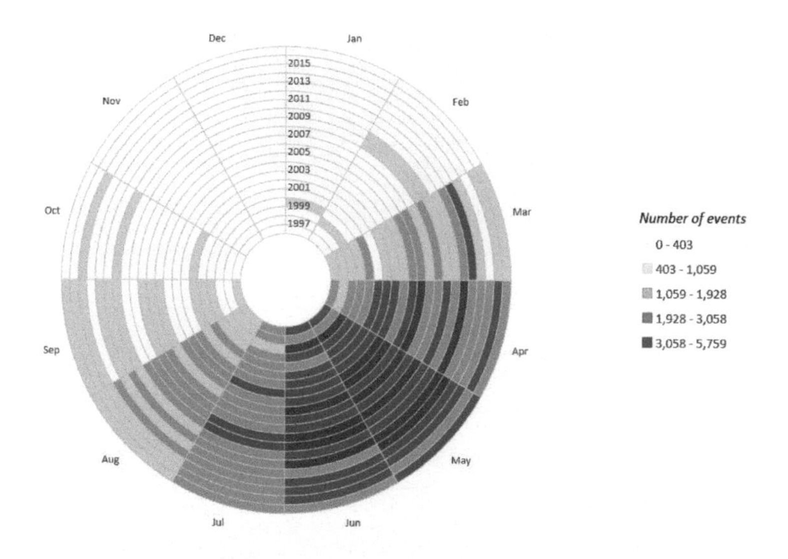

Count of severe hailstorms summarized by months from 1996 to 2016.

The data clock shows two anomalous years in the February wedge that have moderately high counts. In ArcGIS Pro, a data clock links to its related map. Selecting an area on the diagram automatically selects the area on the associated map. Selecting the two anomalous years in the February wedge reveals that these events were

concentrated in southeastern states where anomalous spring-like temperatures can support the development of thunderstorms.

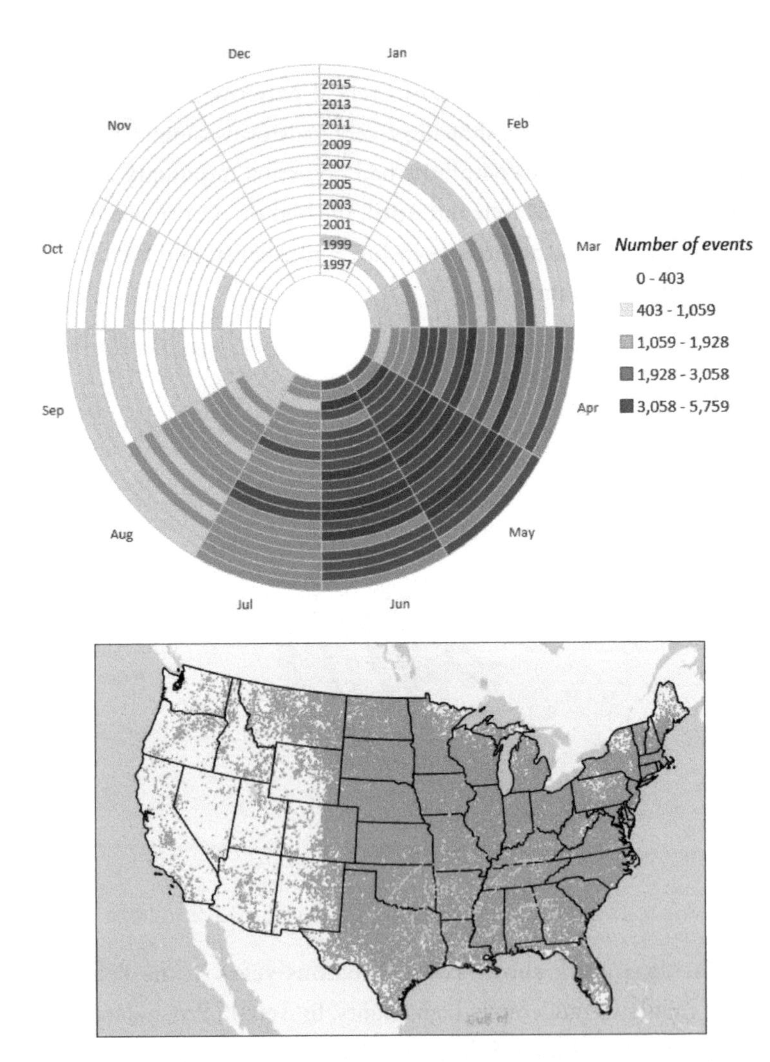

Anomalous hailstorms in February 2008 and 2009. The data clock and map are linked, so selecting the two anomalous years in the data clock (shown in cyan) selects the associated storms in the map.

The Create Space Time Cube by Aggregating Points tool in ArcGIS Pro aggregated more than 272,000 hailstorms into a space-time cube. This tool takes temporal data (hailstorms) and aggregates it into defined geographies (counties in this case). The output of this tool is a space-time cube in which each county has an associated time series.

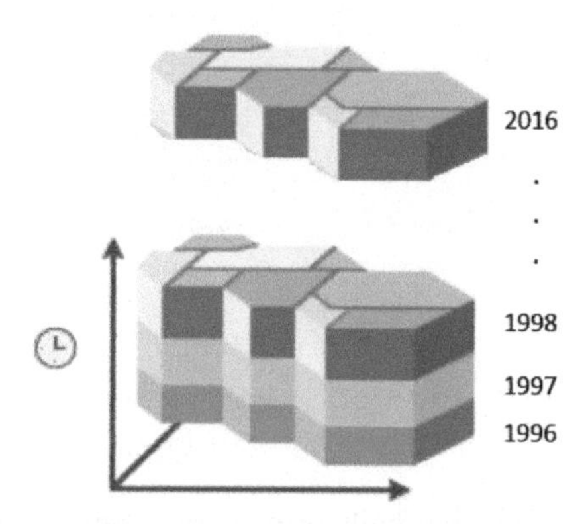

Conceptual view of a portion of a space-time cube in which each location (county) has an associated time series containing the total number of hailstorms for each year (1996 through 2016).

When the cube is created, the temporal trend for each location (county) is captured using the Mann-Kendall statistic. This statistic indicates significant trends (either up or down) in the number of hailstorms for each county from 1996 through 2016. The map of significant trends shows statistically significant upward trends in the upper Great Plains states of North and South Dakota and eastern Wyoming and significant downward trends in northeastern Oklahoma and other southern states.

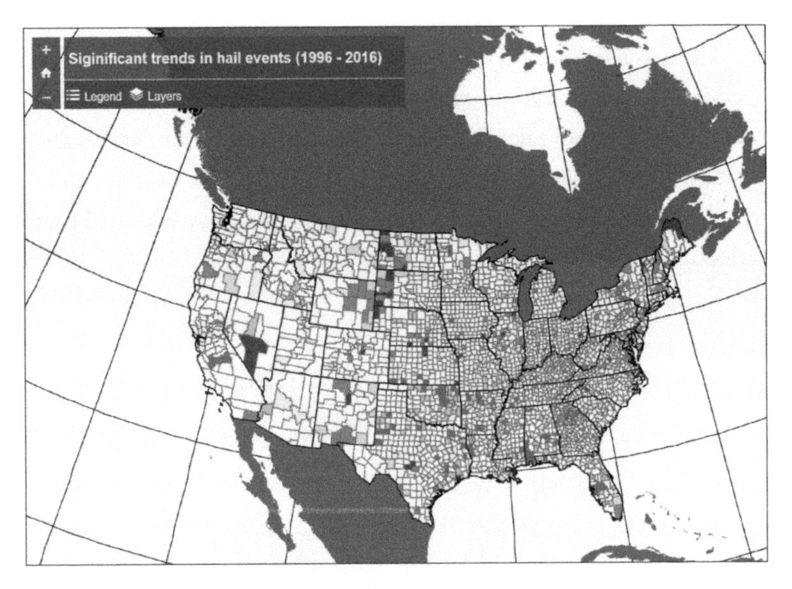

Statistically significant upward and downward trends in time series of the number of hail events from 1996 through 2016 for each US county.

To determine whether hailstorms varied significantly from year to year, the Time Series Clustering tool can partition the collection of time series in a space-time cube, based on the similarity of time series characteristics. Time series can be clustered so they have similar values in time or similar behaviors or profiles across time (increases or decreases at the same points in time). We're interested in the frequency of hailstorms, so the next map (temporal clusters of hail) shows the time series clusters based on the count (the number of hailstorms). From this map and the associated time series graph, we can see three distinct patterns of hailstorms across time in the contiguous United States. The optimal number of clusters, three, was automatically determined by the Time Series Clustering tool using a spectral gap heuristic. Clear patterns exist showing that most of the continent has low hail activity in contrast to the Great Plains. This pattern warrants additional investigation. What is it about the Great Plains that makes more hailstorms happen here?

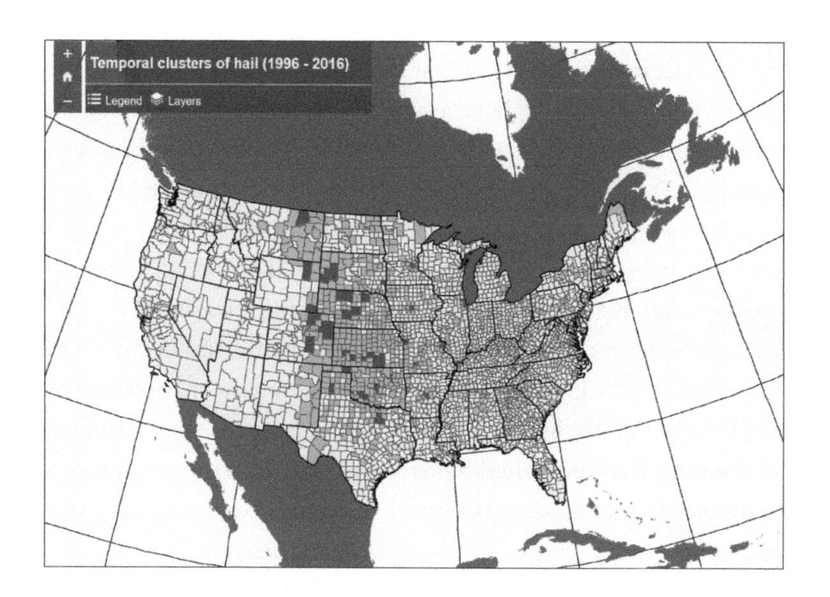

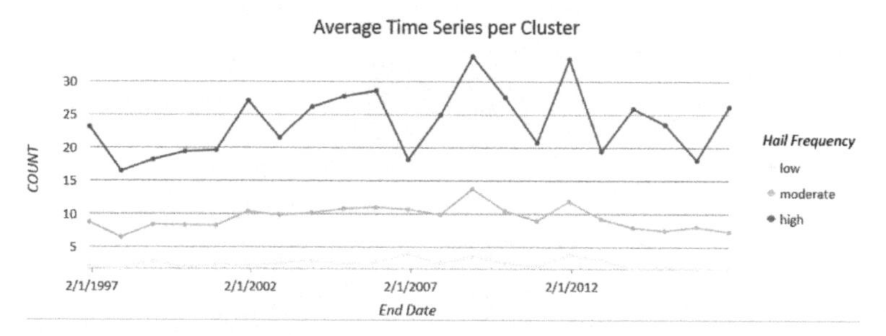

Outputs of the Time Series Clustering tool showing temporal clustering of counts of hail events in each county.

A spatiotemporal perspective

The temporal analysis showed trends in the number of hailstorms in each county from 1996 through 2016. Each location (county) was analyzed separately across time. However, an informative climatology should summarize patterns across space and time. Generating a comparative climatology involves exploring whether clusters of counties have high or low counts of hailstorms and whether temporal trends exist in the intensity of that clustering. Do some regions (counties and their neighbors) have a significantly higher or lower number of hailstorms compared with the overall (contiguous United States) number? The Emerging Hot Spot Analysis tool can answer the question by using the space-time cube created for the previous analysis and identifying clusters as hot and cold spots. The tool identifies any trends (using the Mann-Kendall statistic) in the intensity of the clustering. Exploring the clustering behavior of hailstorms in space and how that clustering changes over time generates a spatiotemporal regionalization. Rather than trying to glean patterns from hundreds of thousands of individual hailstorms, local emergency managers, insurance accessors, or agricultural agents can see meaningful summaries of patterns across space and time concurrently.

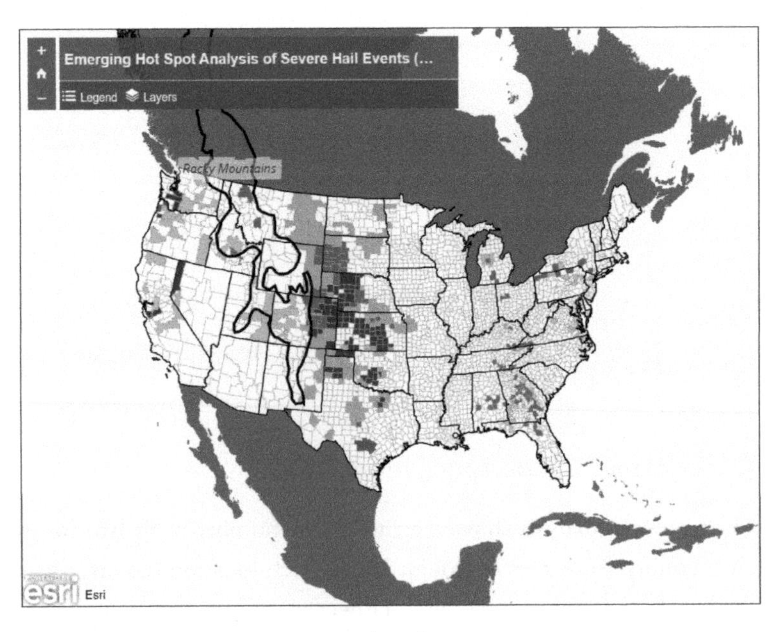

Results of the Emerging Hot Spot tool analysis.

The Emerging Hot Spot Analysis tool categorizes each county based on the type of clustering and trend at that location. For example, several counties in Kansas and nearby areas are categorized as persistent hot spots. In the context of the hail climatology in this study, these counties have been part of clusters of significantly higher number of hailstorms at least 90 percent of the time (19 of the 21 years), with no discernible trend indicating an increase or decrease in the intensity of the clustering over time. In contrast, the counties in western Washington state, northwestern Nevada, and central California are persistent cold spots. These counties have been part of clusters of significantly lower number of hailstorms at least 90 percent of the time (19 of 21 years), again with no discernible trend indicating an increase or decrease in the intensity of the clustering over time.

According to meteorologist Jeff Haby, meteorological factors contribute to the pattern of severe hail across the Great Plains. In part, the high elevation of the Great Plains and a low freezing level (the pressure level where the temperature is freezing) means that hail has less time to melt before it reaches the ground. These two factors plus the tendency for storms to build vertically in this area, thus creating sufficient updraft to carry water vapor above the freezing point, contribute to the high hail frequency.

Integrating climatology and GIS

One of the benefits of a GIS-derived climatology is that geography is an integrative science. It provides opportunities for data integration, spatial analysis, collaboration, and communication of analysis results. The hail climatology can integrate information, such as the amount of land in each county devoted to crops, to calculate a risk map for hail damage to agriculture. The map showing the risk of hail to agriculture shows the relationship between the percentage of time a county was a significant hot spot for hail and the number of acres in agricultural production within a county. The darkest color indicates areas that have high hail and land devoted to agricultural production. This area mimics the extent of the "breadbasket" of the Great Plains. This analysis could repeat using other variables, such as the median age of homes or the percentage of houses without garages, to create risk surfaces, such as risk to roofs or risk to automobiles, respectively.

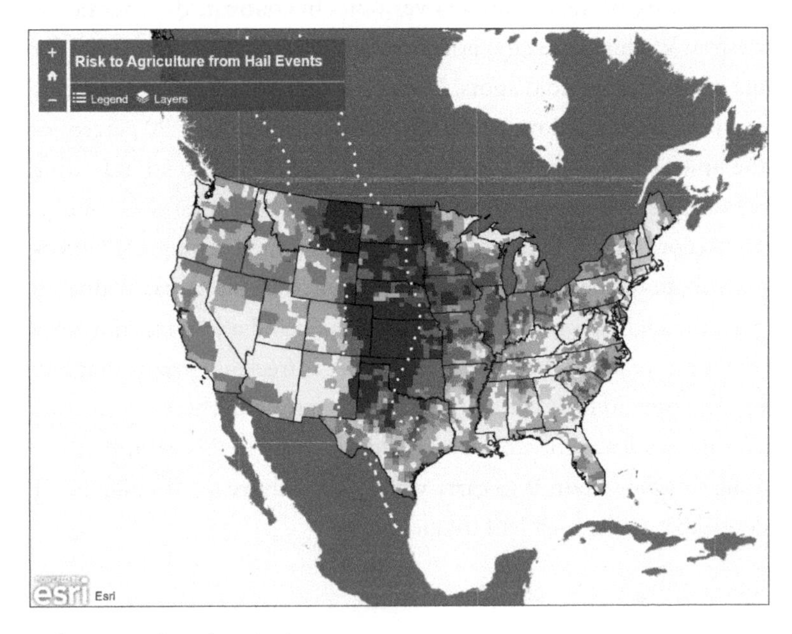

Risk to agriculture from hail events.

A version of this story by Kevin Butler titled "The Science of Where (and When): A GIS-Derived Climatology of Hail" originally appeared on *ArcGIS Blog* on July 30, 2018.

INTEGRATING SEVERE WEATHER DATA HELPS MANAGE SNOW REMOVAL

New York State Department of Transportation

BUFFALO, NEW YORK, GETS SNOW—AND LOTS OF IT. LAKE-effect snowfall often pummels the area, including parts of Erie, Niagara, Chautauqua, and Cattaraugus Counties.

During a snowstorm, Region 5 of the New York State Department of Transportation (NYSDOT) is responsible for maintaining 3,675 lane miles (length plus number of lanes) of highway that commuters, freight haulers, and emergency responders use in western New York. NYSDOT snowplow operators clear roads on their assigned snow and ice beats while emergency managers use GIS to help predict where a storm will make the biggest impact.

With assistance from the National Weather Service, the NYSDOT Emergency Operations Center developed two ArcGIS software-based map applications to help emergency managers better respond to severe snowfall. Practitioners wanted to use GIS to create and analyze reliable snowfall forecast maps to predict a storm's impact on the transportation network.

Getting accurate snowfall forecasts

Finding reliable sources of snowfall forecast data was challenging. NYSDOT does not maintain or generate weather forecasting data or mapping, so researchers at the department explored the data and tools available on ArcGIS Online to see what they could find.

The group discovered that users affiliated with Esri generate mapping products available to the ArcGIS Online user community. One of the maps NYSDOT discovered was the *Severe Weather Web Map*, which contains information from the National Weather Service.

According to federal policy, the National Weather Service must provide a diverse suite of products and services derived from its digital forecast databases. To get the most accurate and up-to-date snowfall forecasts, the National Weather Service uses its National Digital Forecast Database (NDFD), which collects weather data 24 hours a day from field stations across North America. The NDFD makes its weather observations, forecasts, and warnings available to the public via databases that can be used for maps, graphics, and GPS points.

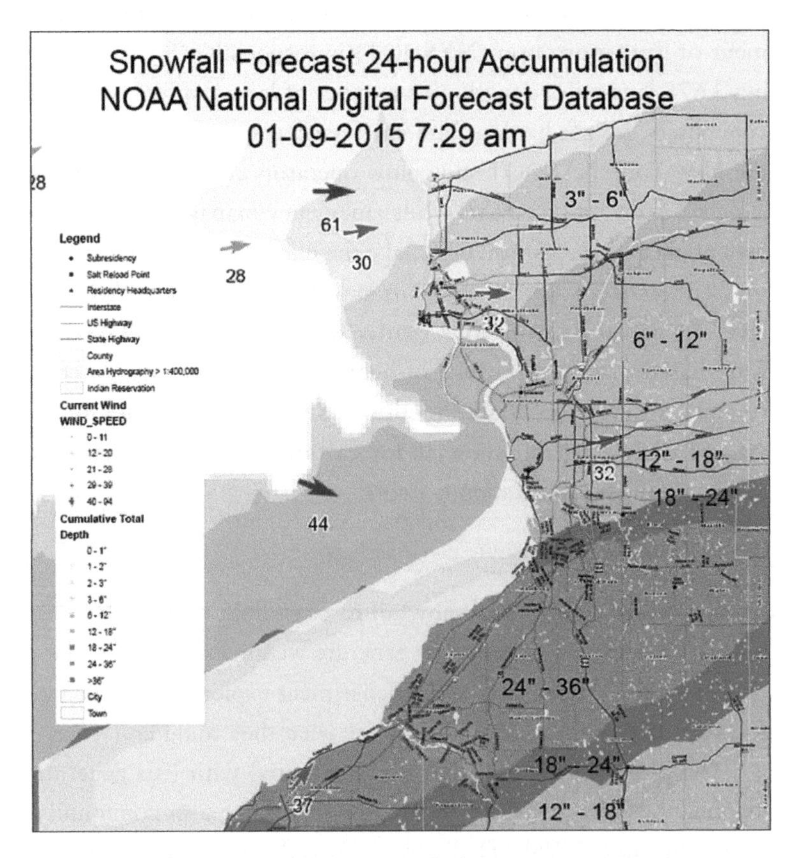

Cumulative snowfall maps show projected cumulative snowfall, wind direction and speed, maintenance areas, and plow routes.

The National Weather Service's Buffalo field office recommended using the *Severe Weather Web Map* for its mapping applications. The map uses NDFD gridded raster data in KML format to show predicted snowfall, so NYSDOT implemented it as its basemap in ArcGIS.

Engaging framework data

NYSDOT also has GIS framework data that consists of feature classes for highways, capital assets, community facilities, and natural landforms. The highway network data, symbolized as lines, documents highways that NYSDOT maintains. Storage facilities that house plowing equipment, salt, liquid deicing agents, and fuel are represented as point feature classes. The framework data displays the maintenance boundaries for each NYSDOT office as well, and Region 5 has GIS data that shows the highway routes traversed by specific plow drivers.

The initial purpose of developing this framework data was to create operational maps for snow and ice removal. However, the NYSDOT researchers repurposed it to figure out how severe snowfall might impact its facilities and snow plow routes. Combined with the GIS data for predicted snowfall, NYSDOT could then understand the magnitude of expected conditions and the spatial context needed to plan snow removal.

Two advantageous maps

After incorporating GIS data into ArcGIS Online, the department created two map layouts that the Emergency Operations Center activated for severe snowfall response. The *72-Hour Cumulative Snowfall Map*, which updates every 12 hours during events to help plan for upcoming operational periods, displays predicted snow accumulation in the area and the parts of the transportation network and locations of NYSDOT facilities expected to be most affected. The *6-Hour Snowfall Forecast Map* goes a step further, displaying the forecasted snowfall amounts in six-hour increments. This allows NYSDOT to estimate the predicted rate of snowfall per hour.

These maps are printed and posted on the walls of the Emergency Operations Center and emailed to affected state transportation offices to inform emergency responders about potential conditions. Managers can also identify areas where a storm may have the greatest impact on transportation, so they can send resources to respond to the snowfall.

A version of this story by Matthew Balling originally appeared in the Fall 2015 issue of *ArcNews*.

WEATHER FORECASTING TAKES A LEAP FORWARD

Weather Decision Technologies

EARTH'S EVER-CHANGING WEATHER PATTERNS CAN BE explained this way: The sun warms the planet unevenly (think of the hot and humid equatorial regions versus the frigid North and South Poles), and that causes weather. The atmosphere redistributes this heat, creating high and low areas of pressure throughout the world. This process results in the development of clouds, winds, precipitation, and widely varying temperatures.

Predicting weather patterns more than a few days in advance can be difficult because of their size and complexity. So, researchers continually refine their instruments and processes to make more accurate forecasts.

Today, with an abundance of satellites and remote sensing devices monitoring weather systems worldwide, meteorologists have more data available to them than ever before. But more data doesn't necessarily translate into improved predictions, which is why Weather Decision Technologies (WDT) uses advanced GIS to better organize and analyze this big data. Esri technology is helping WDT deliver analytic and mapping services to customers.

Better weather surveillance

Weather has always enormously impacted our planet and the atmosphere around it. Monsoons hindered the Mongols from invading Japan in the thirteenth century, and winds played a role when, in 1588, the English defeated the Spanish Armada, since the flotilla could only sail with the wind at its back. In recent years, drought in the western United States has affected food production and prices

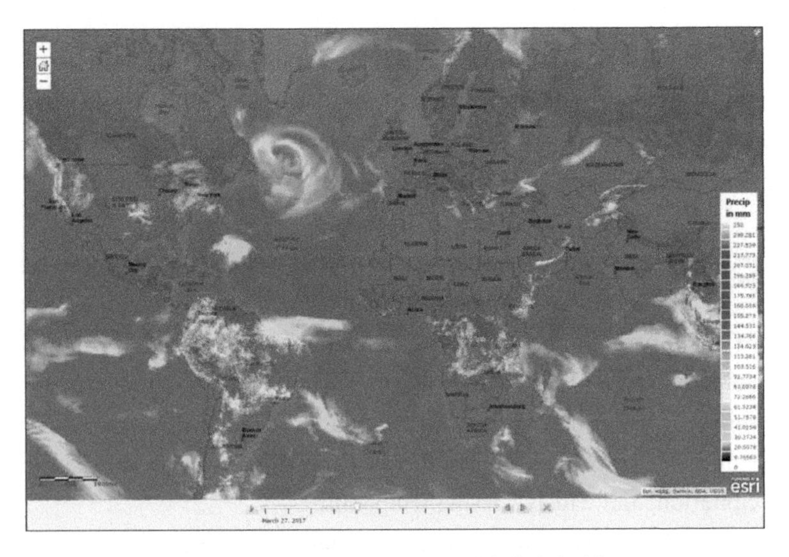

With Weather Decision Technologies' time-enabled global forecasts map service, users can see each day of a 10-day precipitation forecast, such as this one.

nationwide. That is why monitoring the weather has always been imperative.

During the European Renaissance, inventors created instruments to measure local weather phenomena, including temperature, humidity, and atmospheric pressure. When Samuel Morse developed the telegraph in the mid-1800s, local weather observations and measurements were connected in a rudimentary weather surveillance system.

The secretary of the Smithsonian Institution at the time, Joseph Henry, envisioned that the telegraph would "furnish a ready means of warning the more northern and eastern observers to be on the watch for the first appearance of an advancing storm." By the end of 1849, 150 volunteers throughout the United States were helping to make that a reality by regularly reporting weather observations to the Smithsonian. That's when the science of meteorology emerged.

When computer technology was introduced to weather forecasting in 1950, a group of meteorologists at the Institute for Advanced

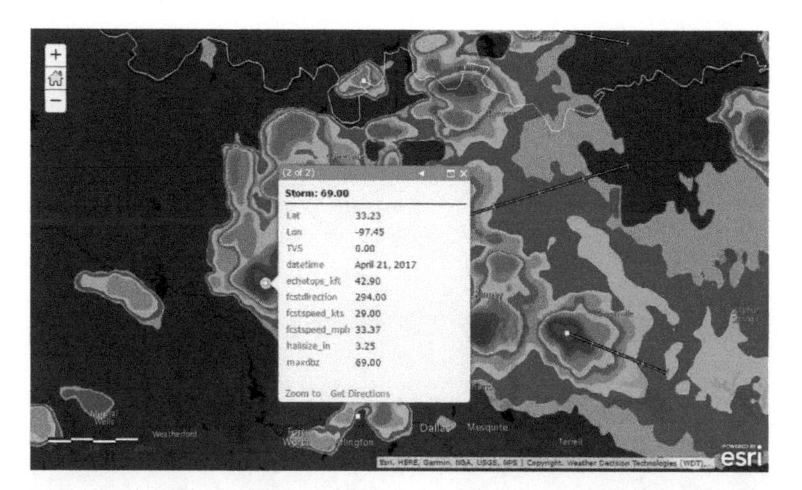

Users can add past, present, and future weather data to their maps and apps to analyze, for example, how the intensity of hail in North Texas contributed to more than $1 billion in losses.

Study in New Jersey produced the first numeric weather prediction, launching the era of modern meteorology and expanded research opportunities.

To better understand developing weather patterns, however, researchers needed satellite imagery. This came in 1964 with the introduction of the Nimbus program. These satellites were designed for meteorological research; the United States launched most of the earth observation satellites during the last 30 years.

Analyzing clouds in the cloud

Data from satellites and remote sensing devices found on land and in the oceans are improving weather forecasting services.

"The amount of data that we collect is enormous, about one terabyte per day," said Matt Gaffner, GIS solutions expert at WDT. "Over the years, we have assembled an archive of almost half a petabyte of weather data."

WDT provides weather forecasting and mapping services to

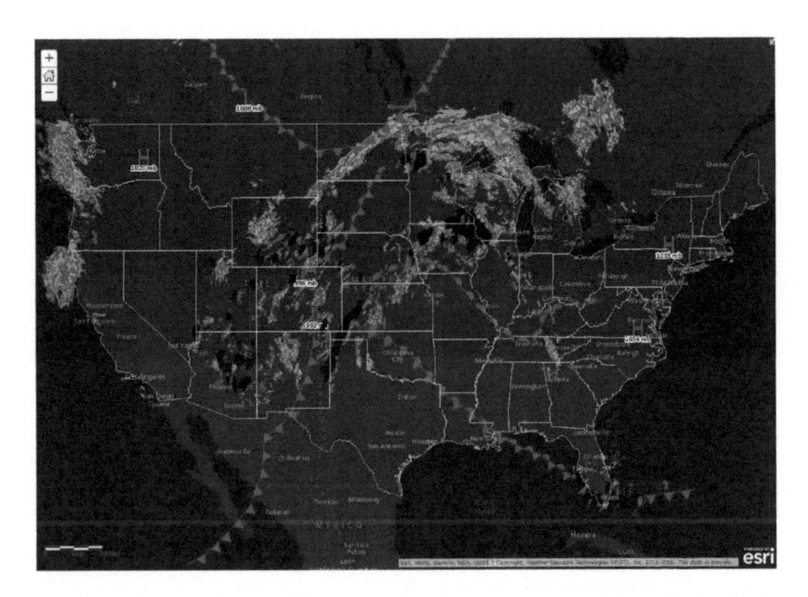

With ArcGIS, users can publish live, dynamic data updated in real time. This interactive map service shows frontal analysis updated every three hours (the purple triangles and red half circles), North American weather radar updated every five minutes (the blue, green, and yellow polygons), and lightning data updated every minutes (the red and black lightning bolt icons).

energy companies and other industries to help them predict electrical outages and keep offshore oil rigs safe. WDT also supports agriculture agencies for crop insurance, freight transportation companies for route design, and concert and sporting events for planning and safety. The company uses ArcGIS to develop its map services for customers.

ArcGIS helps the company publish rapidly updated data and host it as a service for use with other applications," said Gaffner. "This allows our users to quickly add past, present, and future weather data to their maps and apps."

It also enables WDT's partners that build apps for specific vertical markets, such as utilities, to add weather data to their apps using the company's map services.

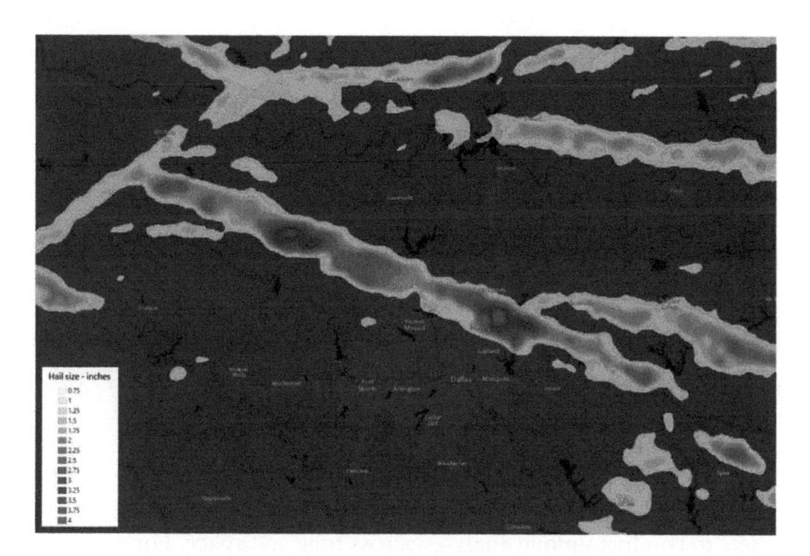

Variation in hail size in North Texas, from light yellow (less than an inch in diameter) to magenta (up to four inches in diameter).

Cloud computing supports WDT's use of Amazon Web Services (AWS) to deploy the analytic and mapping services that ArcGIS provides.

"There are so many advantages available to us by using Amazon Web Services," Gaffner said. "For one thing, it allows us to implement the 'fail faster' mantra. Using the ArcGIS Amazon Machine Image (AMI) capability, we can easily stand up a version of the server in the cloud and try something new—such as using different machine hardware specifications or configuring data services differently—to see if it works or not.

"AWS provides us with reliability because if something goes wrong, we can easily replicate ArcGIS on another machine in the cloud," Gaffner said. "It also provides us with load balancing—that is, we can redistribute the many requests we receive for weather data and map services between our servers. This extra demand normally happens when the weather changes and storms begin to develop. Our datasets increase in size then because there is more radar data accumulating, and our customers need access to that data."

WDT's time-enabled map services allow users to loop through the last 60 minutes of radar data to see where a storm has been and where it is headed. WDT's time-enabled global forecasts map service will provide daily and hourly forecasts for the normal weather variables, such as temperature, precipitation, windspeed, and direction up to 10 days in advance.

Environmental conditions affect business—and decisions

WDT's ever-growing collection of large geospatial datasets allows users to conduct unique analyses across time and space. For example, one of its customers—an oil and gas company—wanted to determine how much the weather affected the productivity of its crews in Oklahoma versus Colorado.

During the summer, it could be 90 degrees in Oklahoma and eastern Colorado, but with more humidity in Oklahoma, the heat index would be higher than in Colorado.

"By using historical weather data," said Gaffner, "we were able to perform this analysis and found that weather can impact crews in two ways: decrease worker efficiency under heat stress and push the heat index above a critical threshold where workers are required to take mandatory breaks."

WDT also considers the impact that the Internet of Things (IoT) can have on geospatial analysis.

"We can provide real-time weather or a weather forecast to your car that might tell you, 'Hey, you should probably stop driving to avoid that storm,' or 'There's a line of thunderstorms moving through your area; you might as well stay at work for another 20 minutes and wait until it passes through and then drive home.'"

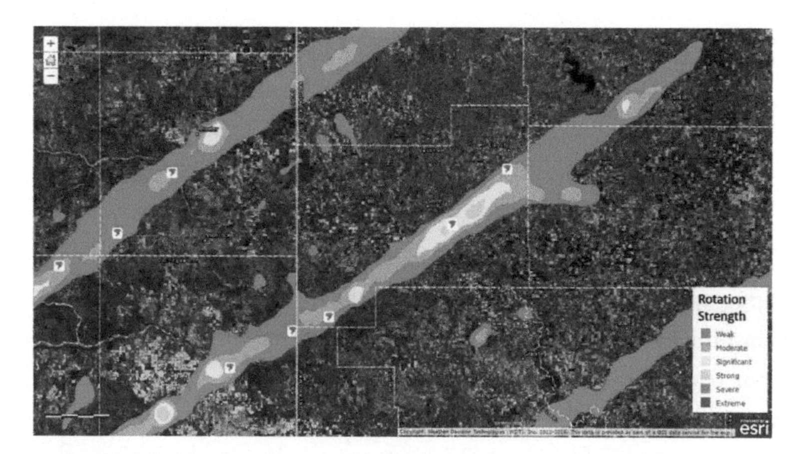

Overlaying a storm's motion vectors (the white boxes with tick marks) onto radar data with storm attributes lets users see where a storm is moving and how fast.

Combining weather and environmental data can help mitigate risk and make more informed decisions. The confluence of big data with smart analysis can save lives and property.

A version of this story titled "Weather Forecasting Takes a Leap Forward with Advanced GIS" originally appeared in the July 2017 issue of *ArcWatch*.

EXTREME HEAT SPURS CLIMATE ACTION

Prague Institute of Planning and Development

THE *PRAGUE DAILY MONITOR* FEATURED THIS HEADLINE IN early July 2015: "Extreme Heat Wave Has Hit Czech Republic." The story said currents of hot air moving from Africa were driving up temperatures across Europe. It was the first of four heat waves to blast the Czech Republic in 2015. More than half of July and August that year recorded extreme temperatures, breaking a record set in the country during a similar string of heat waves in 1994.

Two years after the 2015 heat waves, the city of Prague, the country's capital, issued a document that outlined a four-year plan, beginning in 2020, to "enhance long-term resilience and reduce vulnerability to climate change." To meet these objectives, officials from the Prague Institute of Planning and Development (IPR Prague) use GIS to understand Prague's reaction to climate change now and in the future. Using GIS, IPR Prague can view and analyze the city in its street-level granularity and in its totality.

Prague is vulnerable to extreme heat. Compared with other European cities, it has more paved spaces, built-up areas, and industrial infrastructure—the kind of spaces that can create what are called heat islands. But Prague also contains a significant amount of green space and vegetation, the kind of areas that can offer respite from the heat. From a planning perspective, this tapestry of extremes presents a challenge, a puzzle to solve so that Prague residents can adapt to global warming.

Cities lead the way

As much as coastal communities must address sea level rise, large cities, including landlocked metropolises, face their own challenges because of climate change.

The large conglomerations of people and human development

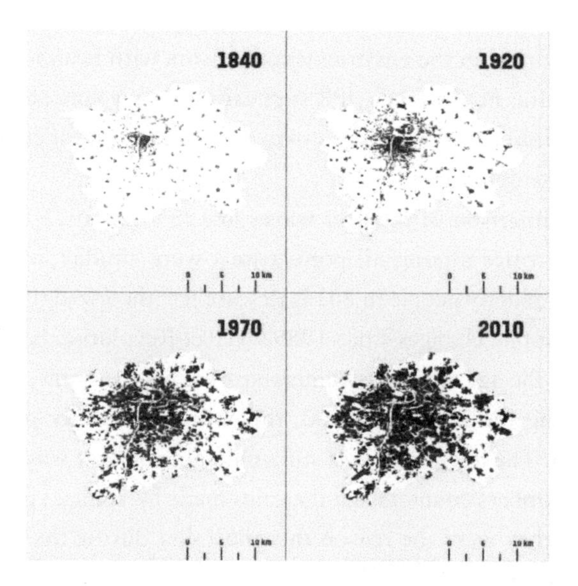

The Spatial Information Section of IPR Prague crafts maps and conducts spatial analysis. This image from its *Do You Know Prague?* brochure shows how the city has grown in size and density over time.

exacerbate the effects of rising temperatures. The economic and social diversity of cities means that certain communities feel the effects more than others, even within one city.

In 2018, the Czech government made climate change mitigation a national priority, a year after Prague released its strategy document.

Prague's plan fulfills the criteria set by the national government but goes further, including making the city carbon neutral by 2050.

Mapping vulnerability

The strategy includes using data in a way that helps IPR Prague understand how climate change currently affects Prague, how those impacts will evolve over time, and how to develop the city to meet these challenges.

Sensors throughout the city measure variables such as temperature fluctuation, solar radiation, and humidity. IPR Prague integrates information from the environmental sensors with health and demographic data. For instance, IPR staff can see heavy concentrations of young children and the elderly—two populations at increased risk from hot temperatures.

A comparison of the heat waves in 1994 and 2015 found that mortality rates among all populations were similar in 1994 but higher for elderly people in 2015. Researchers theorized that positive socioeconomic changes since 1989's Velvet Revolution had left people under the age of 64 less vulnerable to stress over time.

During the 21-year period, the country's elderly population increased. The age group was still vulnerable to heat waves, and its greater numbers counteracted the gains made by younger groups. On balance, they were the reason that mortality during the 2015 heat waves was higher than in 1994.

GIS provides a way to visualize—and therefore contextualize—these statistics. Demographics and other human population data become layers on a smart map. The layers can be set against environmental features of the city, offering graphic representation of how the city and its populations interact.

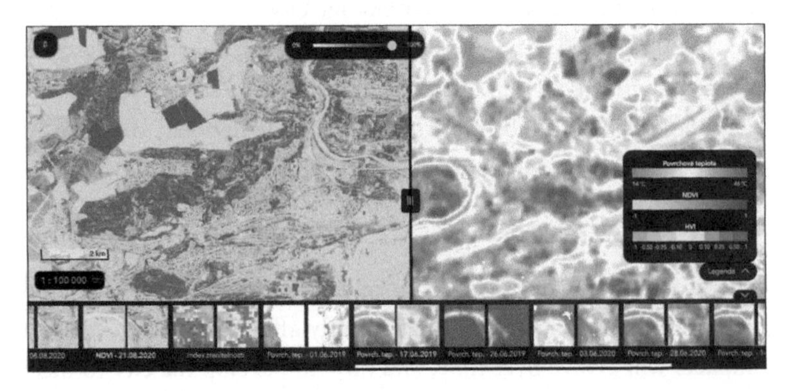

This interactive mapping tool can be used to view vegetation density (*left*) and surface temperature (*right*) to help determine where shade is needed.

This effort grew out of requests from officials for IPR Prague to devise a way to rate the viability of future projects. IPR Prague created a vulnerability index by incorporating data sources through GIS to identify at-risk areas.

Microclimates, macro vision

The next phase of Prague's climate strategy involves using GIS to construct intricate 3D models of the city's microclimates. Once established, these models will provide a way to analyze the effects of mitigation strategies before the city makes any large investment of time and money.

"We'll use them to improve our proposals," said Jiří Čtyroký, director of spatial information at IPR Prague. "For something small, like replanting trees, we won't bother to model it. But if we've got a huge redevelopment project, we'll want to model it right up to the final stage of the proposal." These models also provide a way to communicate these plans, including microclimate modeling, with other government agencies and the public.

A changing data climate

The third tier of Prague's climate strategy involves adding more data sources, working in conjunction with the country's ministry of the environment, joining databases related to environmental indicators, and implementing the strategy as one project.

As the data flow increases, the GIS-enabled map of Prague will become more complex and use more apps. Looking toward Prague's carbon-neutral future, city officials are discussing with IPR Prague the possibility of rooftop photovoltaic and wind-power generation.

A version of this story by Richard Budden titled "Prague: Extreme-Heat Events Spur Climate Action, Using Geospatial Tech" originally appeared in the *Esri Blog* on April 27, 2021.

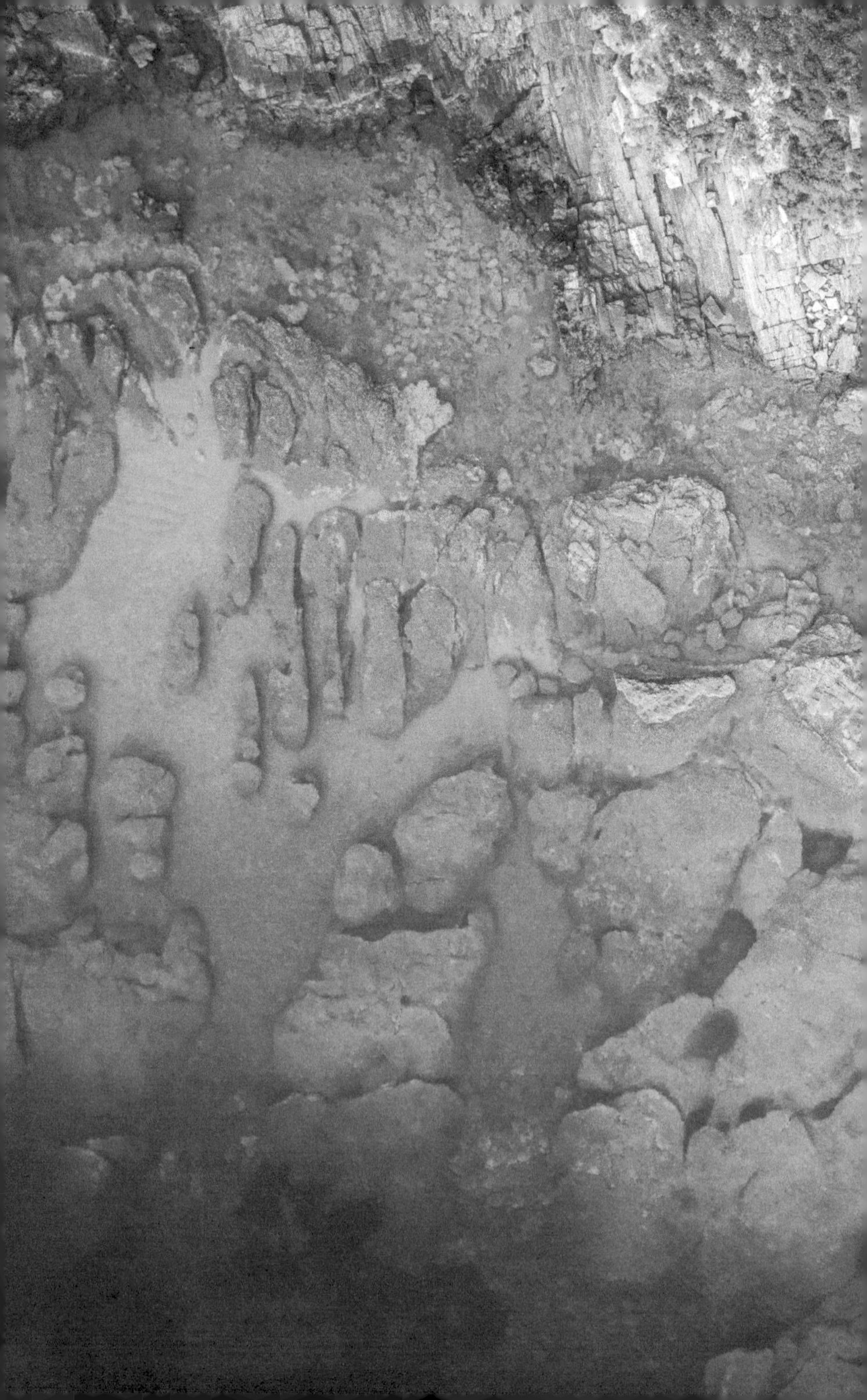

PART 6

MARINE SCIENCE

THE WORLD ECONOMIC FORUM REPORTS THAT IF THE ocean were a nation, it would rank as the seventh-largest globally, with a value of at least US$24 trillion. As climate change impacts the ocean and its economic uses, ArcGIS connects ocean mapping and ocean dynamics. With ArcGIS, marine professionals can share knowledge and plan for the sustainable use of ocean resources for economic growth and improved livelihoods.

Creating a sustainable marine environment

GIS tools are transforming how we map and manage ocean and coastal environments. ArcGIS incorporates AI for mapping and monitoring to measure change in the open ocean, closer to the shore, and coastlines. Marine scientists aim to achieve a sustainable ocean environment. They use statistical functions, geospatial tools, and predictive analytics with time-enabled 3D visualization to understand ocean characteristics:

- **Rediscover the seafloor and marine habitat:** Adopt a holistic view of the marine environment through interactive maps and dashboards.

- **Get secure marine data collections:** Access and integrate disparate ocean data in secure cloud or hybrid data services to facilitate data exchange.

- **Apply ready-to-analyze data sources:** Combine out-of-the-box modeling environments with open-source spatial algorithms, open data libraries, and programming languages.

- **See underwater patterns more clearly:** Increase marine understanding using 3D mapping. Explore the ocean depths with subsurface profiling, 3D simulations, time series animations, and interactive maps to reveal patterns.

- **Restore, preserve, and protect the ocean:** Develop marine ecosystem planning scenarios through suitability modeling.

- **Operate securely with other scientists:** Access marine layers across systems and share research with colleagues and other teams using a built-in collaborative framework.

GIS in action

Next, we'll look at some real-life stories of how organizations are using GIS to provide a new window to the undersea world.

ACHIEVING SUSTAINABLE PROSPERITY WITH SEAFLOOR MAPS

Clearwater Seafoods

WHEN EXECUTIVES FROM CLEARWATER SEAFOODS MET with an ocean mapping expert, it opened a new window to the undersea world. "It just blew us away because all we'd ever had were surface charts of the ocean," said Jim Mosher, Clearwater's director of harvest science. "Now, all of a sudden, we could see the contour and shape of the floor. It was a showstopper."

The Nova Scotia-based brand specializes in luxury seafood, including scallops, clams, crab, shrimp, and lobster, that it gathers mainly from the seafloor in coastal Canada. Sonar imaging and 3D seafloor models have shifted the brand's entire outlook.

Some of Clearwater's catch, such as clams, live on the seafloor, creating homes in the silt and mud known as the substrate. Clearwater must understand when and where to harvest and not to harvest clams and other shellfish to protect the resource and the environment. Seafloor models that are created using GIS improve harvest operations and sustainability.

"Rather than the traditional method of just going to sea, we do a lot of pre-trip planning," Mosher said. "We've moved the operation from one that senses and responds to one that predicts and then acts, targeting fishing sites with high precision. Our entire mentality has changed with the increased availability of robust data and mapping over the years."

Tales of topographic oceans

After meeting with the mapping expert, Mosher's team began investing heavily in technology to create bathymetric maps of habitats—

topographic seafloor maps with detailed depth contours to show the size, shape, and distribution of underwater features. Then they mapped sediment types.

"We knew that some of these species that are filter feeding, like scallops, don't like living on sand because they choke, so they aggregate on gravel," Mosher said. "All of a sudden, we were targeting gravel patches. But what we were doing was still really crude compared to what we have today."

Mosher's team continued to gather data on seabed types, rock structures, and other oceanographic ecosystem conditions, which function as layers on a basemap. The layers can be combined to gain an ever-growing location intelligence. Bathymetric information can be overlaid with fishing practices to determine best practices.

Before setting sail

Clearwater's precision approach is a matter of corporate responsibility. In Canada, companies are granted rights to a specific fishery to hunt specific species; Clearwater sees it as its mission to steward the ocean and protect its holdings through responsible fishing practices.

When a Clearwater ship leaves port to gather a catch, the procedure varies by species. To fish for scallops, a ship measuring between 140 and 150 feet carries a crew of 16.

Scallop gathering often includes an extensive preplanning expedition with a survey ship in collaboration with other industry stakeholders. Clearwater identifies a potential fishing area and divides it into individual grid cells on a smart map, using ArcGIS Pro. An underwater camera captures seafloor images that assist in understanding species composition.

"We digitize the images, count the scallops in the grids, and complete analytics to generate density models—and we map all of it," Mosher explained. For clam operations, captains will rely on

Fundy Leader serves as a scallop harvesting vessel and Clearwater's research vessel, conducting video surveys and rake surveys to assess scallop bank conditions.

backscatter maps that use multibeam technology to identify substrate hardness. These maps help Clearwater understand biomass in specific areas and calculate how quickly an area will rebound.

Clearwater partners with academic institutions in Nova Scotia to create thematic maps. "Geology, sediment, backscatter—and on top of that we add our survey layers," Mosher said. "And then of course we have catch data on maps that show exactly where the ship has been in the past, so we don't go and fish in the place we were last."

Before a fishing expedition, a ship's captain will study these map layers to make plans. The data will travel onboard because each ship is equipped with ArcGIS Pro.

A vertically integrated company that processes and distributes its catch, Clearwater outfits its ships to be floating factories. Their GIS helps Clearwater see and organize information related to every facet of harvesting and sustaining shellfish populations. It allows Clearwater to further its goals related to sustainability, create a record of its activities for food safety, and convey the origin of each catch. In this way, GIS becomes a 4D tool, cataloging the space and time for the present catch and all past catches.

A changing ethic

Extensive mapping can yield immediate gains. However, Clearwater leaders consider the time and money their teams spend on mapping as a long-term investment, allowing the company to adopt a sustainable business model known as *blue growth* instead of a short-sighted plan that aims for immediate gains.

With increasing amounts of high-quality data, Clearwater teams use their knowledge to forecast and predict harvest outcomes. They use sophisticated assessment models that determine stages of growth and log activities to avoid returning to the same fishing grounds until appropriate time has passed. Clearwater also considers weather influences, species biology, and other relevant details to determine the most effective methods of harvesting.

This approach improves success rates and efficiency. Clearwater spends less time finding a catch, which reduces the carbon footprint and operational expenses. The company's sustainability efforts have contributed toward maintaining its Marine Stewardship Council certification—a coveted badge of sustainability—for adhering to strict environmental and production standards.

"We're really moving the operation from just considering ourselves as hunters to being one of culturing and nurturing," Mosher explained. "To be able to look at the seafloor and understand population characteristics—these are the tools we need to do that. It supports the objectives of sustainability our company has."

A version of this story by Corey Nelson titled "Canadian Seafood Company Achieves Sustainable Prosperity with Seafloor Maps" originally appeared in the *Esri Blog* on May 4, 2021.

3D AND LOCATION INTELLIGENCE HELP DRIVE A SUSTAINABLE OCEAN ECONOMY

Esri

WITH ASSETS VALUED AT MORE THAN US$25 TRILLION, including a $2.5 trillion annual gross domestic product (GDP), the ocean economy also requires conservation. And no wonder: the ocean can protect us from climate change, absorbing 25 percent of the world's climate emissions and 90 percent of the heat caused by those emissions. It generates 50 percent of the oxygen and is the largest biosphere on the planet, home to 80 percent of life on earth.

Business leaders and companies around the world increasingly understand the need to embrace strategies and practices that support a sustainable ocean economy. Technologies such as 3D, digital maps, and location intelligence are supporting the global challenge of preserving the ocean as its economic use increases and climate impacts intensify.

The ocean suffers from overfishing, habitat destruction, pollution, biodiversity loss, and climate change even as the ocean protects people from the consequences of the economic activity it supports.

The ocean serves as a dual solution for climate change. In addition to all its life-sustaining forces, it provides renewable energy and sustainable seafood.

"For the first time, our knowledge of the ocean can approach our knowledge of the land," said Dawn Wright, geographer, oceanographer, and chief scientist at Esri. "We can turn the unknown deep into the known deep."

The release of the world's first complete 3D map of the ocean supports innovation in ocean-related data and sustainability solutions.

"Seeing the ocean in its true depth and complexity is exactly what we need," Wright said. "It's what we need if we hope to reduce the risk of critically damaging or exhausting marine resources, if we hope to preserve the world's fisheries, or to anticipate when a warm current will turn into a devastating hurricane. It's what we need if we hope to tackle the growing continents of plastic, wastes, and other pollutants threatening marine life."

Ocean economy and innovation

The ocean teems with its own life, but its health impacts all human life. Recognizing this, a growing cohort of socially responsible companies have started building sustainability solutions to match business opportunities—achieving profit while preserving the ocean.

Their work often revolves around complex and real-time data, stored and processed with a GIS and visualized on digital smart maps. Location intelligence helps leaders in shipping, energy, logistics, and fishing industries answer these and other questions:

- Where would offshore wind turbines have the least impact on commercial fishing?
- Where should a new transatlantic submarine communications cable go to avoid interference with scallop beds, rare deep-sea coral habitat, or sand mining areas needed for beach restoration?
- Where are appropriate areas for ships to transit in the Arctic (now that it's no longer covered with ice year-round) to minimize impacts on sensitive ecosystems?

Companies that approach sustainable seafood by increasing supply chain traceability will help protect the ocean while meeting the growing consumer demand for ethical practices.

Taylor Shellfish Farms, for instance, maintains a sustainable system that begins in its hatcheries in the state of Washington where shellfish are bred and the tidal beaches where they reach maturity—and extends through harvesting, processing, and distribution. Every link in that process must meet sustainability standards set by the company and industry regulators.

GIS-based maps and dashboards provide visibility into every aspect of the company. Taylor leaders use their GIS data and reporting to maintain responsible environmental stewardship. At the same time, the technology supports process improvements that make it a profitable business enterprise. The efficiency takes many forms, from mobile apps that let farmers update information in the field to smart maps that show where certain techniques or environmental conditions are bringing stronger yields.

Smart shipping, sustainable shipping

Commercial ships produce an amount of carbon—800 million tons per year—that exceeds the output of most countries. Shipping companies are working to decrease greenhouse gas emissions by designing more efficient vessels, a move that furthers the sustainability cause while reducing business expense.

Location intelligence helps shipping companies become more efficient.

Individual ships can track operations using GIS, providing a view of how onboard systems are functioning. For the larger project of fleet management, GIS helps operators track emissions using smart maps to report back to international regulatory agencies.

Taking a circular approach to the blue economy

Even in industries such as retail and manufacturing, socially responsible companies aim to extend sustainable practices throughout a project's life cycle. This often means enacting steps to reclaim or recycle materials after the product has completed its original use.

The circular economy is a way for companies that exist far afield from the blue economy to support ocean sustainability. A circular approach to manufacturing lowers costs while decreasing the amount of plastics that reach the ocean. Likewise, precision agricultural techniques increase efficiency in food production while limiting the amount of pesticides, sediments, and organic matter that pollute the ocean. These practices support the blue economy, which the World Bank describes as the "sustainable use of ocean resources for economic growth, improved livelihoods, and jobs while preserving the health of ocean ecosystems."

Now is the moment

In launching its "Decade of Ocean Science for Sustainable Development" in 2021, the United Nations called for "a healthy and resilient, safe and productive ocean" and international cooperation among all sectors and communities.

"We need the capacity and competence of the business community to solve this challenge," said Lise Kingo, former executive director and CEO of the UN Global Compact. "Ensuring a healthy marine environment is not only necessary for many ocean companies to continue to operate in the long term—innovating and investing in new ocean solutions also provides a significant business opportunity."

A version of this story by Marianna Kantor titled "How 3D, Location Intelligence Can Help Drive a Sustainable Ocean Economy" originally appeared on Forbes.com on February 24, 2021.

SUSTAINABLY GROWING THE BLUE ECONOMY THROUGH SHARED OCEAN INFORMATION

Esri

THE GLOBAL COMMUNITY INCREASINGLY STRIVES TO preserve a healthy ocean while also relying on it as an economic engine. The World Bank has described the potential for new and innovative uses of the ocean with high economic value; however, those uses also threaten the marine ecosystem. The UN has called for a global consensus to incorporate environmental and social dimensions for a sustainable ocean.

One of the UN's 17 Sustainable Development Goals (SDGs) adopted in 2015 is ocean-specific: "Conserve and sustainably use the oceans, seas and marine resources for sustainable development."

In 2017, 189 intergovernmental organizations gathered to confer about how to achieve this goal, with over 1,400 commitments made. They agreed on the need for spatially explicit and publicly available information about coastal and ocean areas. Esri's vision to create products that blend data and geography can turn this consensus into reality.

Using GIS, organizations today can more easily collect and display data about human impacts on the ocean through spatial modeling, predictive analysis, and other emerging technologies.

In the past, disparate government agencies, academic institutions, private companies, and conservation organizations collected much of the spatially detailed data on ocean ecological resources and human uses. Obstacles included the lack of collaboration in their approaches to permitting and planning.

The siloed approach created challenges for permitting agencies to consider the marine environment as a whole. The lack of a common vision for the use of ocean space resulted in inefficiencies and lost opportunities to avoid conflict.

The use of GIS creates opportunities for a more integrated approach. Data portals established for the Mid-Atlantic, Pacific Islands, Caribbean, and other regions offer verified and publicly available data and maps. Sources ranging from state and federal agencies to private industry and sovereign tribal nations have contributed information about fishing grounds, whale migration routes, ship traffic, recreational boating, and more.

The portals identify data gaps to fill. Esri's mapping tools are components of each of the data portals, ensuring that data layers can be visualized to drive decisions.

These data portals help underpin ocean planning. Safeguarding and improving the health of the ocean and its ecological resources depends on understanding where and how the ocean is used.

Public data portals support a sustainable blue economy

In the Northeast United States, accessible and comprehensive data allowed for the development of the first shellfish farm in Atlantic Ocean federal waters. The Northeastern Massachusetts Aquaculture Center (NEMAC) at Salem State University used the Northeast Ocean Data Portal to define an area for growing blue mussels that will not harm protected whale populations. NEMAC also evaluated maps including fishing activity, vessel traffic, fishing closures, and essential fish habitat in the area to reduce conflicts.

Not-for-profits, ports, cities, states, regions, and countries are growing the blue economy

Globally, efforts are underway to jumpstart ocean-related sustainable economic development. Many of the economic opportunities rely on ocean data and planning. Some innovation centers include industries that are collecting data using underwater robotics. Many of these initiatives include a physical place where businesses, scientists, and educators address ocean issues.

At the Port of Los Angeles, for example, AltaSea is a 35-acre campus designed to expand science-based understanding of the ocean, start and sustain ocean-related business, and develop ocean-related education. Their initial focus areas are sustainable aquaculture and blue tech (underwater robotics).

In Reykjavík, Iceland, the Iceland Ocean Cluster is connecting entrepreneurs and businesses in the marine industries. Its facility in Reykjavík supports entrepreneurs, startups, and businesses in the marine industry. Its work includes promoting fisheries products for new markets such as pharmaceuticals.

In other cases, innovation clusters have coordinated and convened virtually:

- The Maritime Alliance is an industry association for the BlueTech cluster in San Diego, California, which focuses on business ecosystem, economic, and workforce development by bringing together academia, industry, and governments.
- The Alaska Ocean Cluster Initiative strives to grow Alaska's blue economy by uniting industry, academia, nonprofit organizations, and governments in a growth strategy.

The future is blue

The emergence of ocean innovation centers and clusters indicates the worldwide interest in growing a sustainable blue economy. The UN's ocean planning process provides a venue for businesses, recreational users, and conservation groups to come together with tribes and state and federal agencies to define a common vision for shared ocean spaces, lessen conflicts, and support ocean health.

A version of this story by Sandra Whitehouse originally appeared in the *Esri Blog* on December 14, 2017.

A GLOBAL APPROACH TO PREVENTING PLASTIC FROM REACHING THE OCEAN

Namma Beach, Namma Chennai

LIKE MANY COASTAL AREAS AROUND THE WORLD TODAY, the beaches in Chennai, India, attract enormous amounts of plastic debris. For teenage surfer and local resident Karan Chakravarthy, the presence of plastic at his favorite surf spots was distressing. So he decided to do something about it.

Chakravarthy joined other volunteers to collect trash with a nonprofit called Namma Beach, Namma Chennai (which translates to "Our Beach, Our Chennai"). In 2021, the organization removed 176,000 pounds of plastic waste from Chennai's beaches. But Chakravarthy felt that more could be done.

He contacted his grandfather, Mandyam Venkatesh, who lives in San Diego, California, and obtained a $5,000 grant from Venkatesh's Sunrise Rotary Club to further support Namma Beach, Namma Chennai. Through his grandfather's Rotary connections, Chakravarthy also met Carl Nettleton, the founder of OpenOceans Global, a San Diego-based organization that employs geospatial technology and community science to help stop the flow of plastic into the world's oceans.

Nettleton gave Chakravarthy a Survey123 form that he used to record information about beaches in Chennai that are littered with plastic. The data was then uploaded to the OpenOceans Global geospatial portal. The organization's web-based *Ocean Plastic Map* depicts a red bull's-eye symbol on India's southeastern coast (and on dozens of beaches around the world). Clicking the icon displays a pop-up that includes information about the plastic waste found on Chennai's beaches, including where it comes from and what is being done to clean it up.

Karan Chakravarthy used ArcGIS Survey123 to record data about beaches in Chennai, India, which are consistently littered with plastic. Photo courtesy of Karan Chakravarthy.

Nettleton hopes that community scientists will do what Chakravarthy has done and record data for OpenOceans Global about beaches that are fouled by plastic trash, with GIS users taking the lead.

How plastic waste gets to the ocean

Eleven million metric tons of plastic reach the ocean each year, and that number could triple by 2040 if large-scale solutions aren't developed quickly, according to research by The Pew Charitable Trusts and sustainability consultancy SYSTEMIQ.

"The common perception is that most ocean plastic is in the Great Pacific Garbage Patch, which is estimated to be twice the size

Volunteers cleared 176,000 pounds of plastic waste from the beaches in Chennai, India, in 2021.

of Texas," Nettleton said, referring to the largest of five garbage patches in the world's oceans.

However, a Florida State University study published in *Frontiers in Marine Science* found that about 75 percent of mismanaged plastic waste turned up on beaches from 2010 to 2019.

"Plastic ends up on shorelines because the majority of ocean plastic comes from land, and most of that comes from rivers," Nettleton said.

OpenOceans Global seeks to identify how plastic flows into the ocean and accumulates on those shorelines. According to a study funded by the nonprofit The Ocean Cleanup and published in *Science Advances*, about 80 percent of plastic that ends up in oceans comes from more than 1,000 rivers—many of which are in Asia, Latin America, and Africa. Researchers found that small urban rivers in places with poor trash management practices convey the most plastic pollution to the ocean. But this doesn't mean that the trash necessarily originates there. Countries with upper-income economies—such as the United States, Japan, and France—outpace the rest of the world in plastic consumption and then ship more than a million tons of recyclable plastic overseas each year, often to places with trash management issues.

"We think there are ways to stop plastic waste from reaching the ocean if we know where it comes from geographically," Nettleton said. "Even though the United States and other developed nations produce most of the plastic, the Florida State study found that 55 percent of ocean plastic reaches the ocean from five countries: China, the Philippines, India, Brazil, and Indonesia. If the Philippines sends almost 16 percent of the world's plastic to the ocean via its rivers, as this study discovered, the world could focus on developing solutions for this one country and bring global resources behind it to get the Philippines as close to a zero ocean plastic contribution as possible. We could see which solutions work best there—whether it's implementing river intervention technologies to stop the plastic from reaching the ocean, developing new products to replace plastic, or implementing new processes for trash management—and then replicate those models in other high-plastic polluting countries."

A global view of where plastic pollution originates

The project started with OpenOceans Global using ArcGIS Online and ArcGIS Living Atlas of the World to develop a map that focuses on where plastic litters the world's coastlines.

"You can click on the map and see the rivers of the world, major ocean currents, and a highly detailed point-in-time snapshot of ocean currents," Nettleton said. "These tools help people better understand how plastic debris travels."

Map users can activate layers that show the top 20 rivers that contribute plastic to the ocean and where plastic collects in ocean gyres. They can also see the survey data that community scientists contribute about plastic pollution on their local beaches.

Using Survey123 on their mobile devices or desktop computers, they enter the beach or coastal area's name, pinpoint its location on a map, upload an image that shows the waste accumulation, provide a description of the issue, predict where the trash is coming from, and record what is being done to solve the problem. They also enter their contact details and information about organizations they work with.

After an entry is submitted, a temporary red dot symbol automatically appears on the OpenOceans Global web map. A team at the organization verifies the information and, if it all checks out, turns the red dot into a red bull's-eye, indicating that the coastal area is pervasively fouled by plastic.

"The way plastic has been approached is as a local problem—you know, 'my beach has plastic on it, so I'd better not use plastic straws or plastic bags anymore,'" Nettleton said. "Well, that's important. But there isn't yet a global view of where that plastic comes from."

The more entries that are contributed via OpenOceans Global's Survey123 form, the clearer this global picture will be. When OpenOceans Global has enough data points, the team can distinguish the sources of plastic pollution on specific beaches—from rivers, stormwater systems, or inadequate local trash management—and add new map symbology to reflect that.

"Knowing where the plastic originates helps identify solutions to prevent plastic from reaching the ocean," Nettleton said. "For instance, placing barriers in small, local rivers can capture trash before it gets to the ocean. Theoretically, that will reduce the amount of plastic that ends up on beaches, and at a certain point, those beaches won't be pervasively fouled by plastic anymore."

Tracing plastic through the open ocean

Identifying the source of trash is a challenge when it arrives onshore through the open ocean. In the Galápagos Islands, whose once-pristine coastlines gather plastic waste, an international research

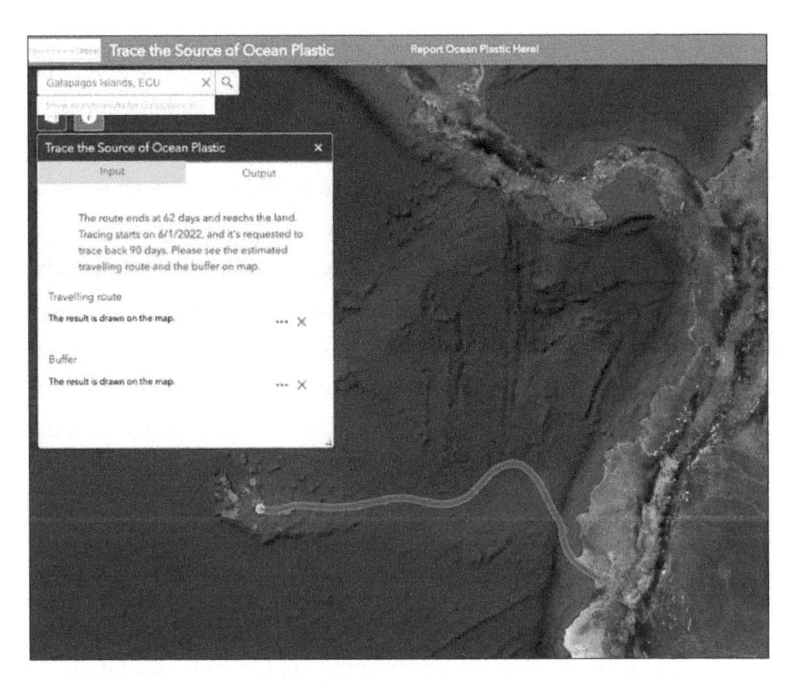

A tracing app tracks plastic found on coastlines back to its source.

initiative called Plastic Pollution Free Galápagos uses a forensic process to analyze plastic debris and determine its source. The initiative's research suggests that more than 60 percent of plastic that ends up in the Galápagos Islands stems from mainland South America (mainly southern Ecuador and northern Peru), with about 30 percent coming from nearby fishing vessels and less than 10 percent from local towns.

Not every coastal area has access to forensics data. So the team at OpenOceans Global worked with a student working toward a master's degree at the University of Redlands to create a mapping app that traces plastic on coastlines back to its source.

The app uses Ocean Surface Current Analysis Real-time (OSCAR) data, which shows surface-level ocean currents, along with ocean current data from NOAA and satellite data from NASA. App users can click an area of the ocean immediately adjacent to where coastal plastic was identified, and the app will create a route to its source. For the Galápagos Islands, the OpenOceans Global tracing app aligns with Plastic Pollution Free Galápagos' forensics research.

"As solutions are put in place and shorelines are no longer fouled by plastic, we will turn the icons on our map green to show that the problem has been fixed," Nettleton. "That's the ultimate goal."

A version of this story originally appeared in the Fall 2022 issue of *ArcNews*.

NEXT STEPS

A geographic approach to earth sciences

EARTH SCIENCES PROFESSIONALS USE LOCATION-BASED data to find, share, and analyze information when they need it. Using a geographic approach with ArcGIS technology, researchers from across different organizations can more easily collaborate on equitable solutions, visualize data to reveal patterns, and communicate their findings across wide audiences.

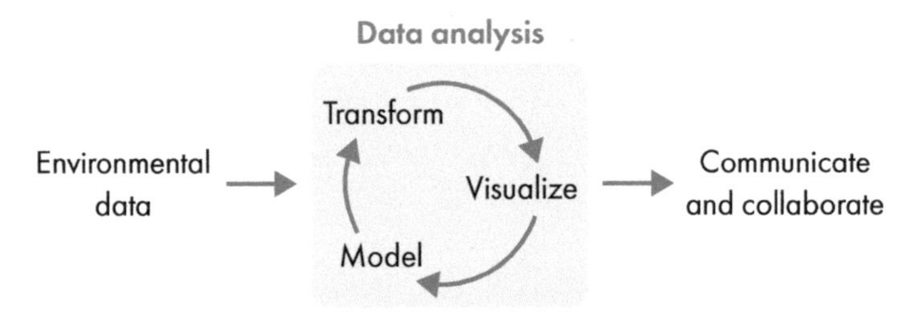

ArcGIS is a collaborative tool for open science and its application.

ArcGIS is fundamentally an open system: Esri aims to ensure that ArcGIS software is interoperable with the technologies that different organizations use to achieve their interoperability goals.

ArcGIS technology supports open standards and industry data standards: Esri supports and contributes to more than 200 open-source projects while delivering more than 350 of its own. Many projects are science-based, and ArcGIS software provides direct read, import, and export for over 300 data formats, with more than 200 for images and sensors.

ArcGIS software furthers collaboration: Users can use an ArcGIS open-data extension to collaborate with others through web services. Esri provides access to scores of APIs and SDKs for open science and has many open-source tools in GitHub.

ArcGIS architecture is extendable: Users can add to it and embed it within other systems. Thousands of companies and organizations use ArcGIS in complex enterprise environments.

By using ArcGIS, science professionals spend more time on analysis, discovery, and innovation. It provides access to the foremost collection of global spatial information, streamlines daily workflows, helps uncover trends, and improves understanding and decision-making.

- **Access big data:** ArcGIS supports the use of big data. Users can manage and aggregate disparate information in a unified infrastructure for secure and timely data exchange with stakeholders.

- **Bring clarity to collections of data:** ArcGIS helps users integrate out-of-the-box modeling environments with open-source spatial algorithms, data libraries, and programming languages. It supports research with ready-to-analyze and streaming data sources.

- **Data analysis:** A comprehensive set of analytic methods and spatial algorithms helps users make connections between disparate data. ArcGIS improves predictive modeling in a competitive environment.

- **Collaborate with the GIS community:** ArcGIS provides a more inclusive approach to science and innovation with a framework that helps engage departments, science disciplines, and the public.

- **Communicate science:** ArcGIS allows users to share their work with anyone, anywhere. Users can more easily work together on analysis projects, share analysis workflows, communicate results, and enable others to perform the same analysis tasks to make informed decisions.

For links to examples shown in this book and additional resources, visit the book web page.

Implement your GIS

If your organization is not already using ArcGIS, resources are available to help you get started.

ArcGIS Online

With a subscription to ArcGIS Online, organizations can manage their geographic content in a secure, cloud-based environment. Members of an organization can use maps to explore data, create and share maps and apps, and publish their data as hosted web layers. Administrators can customize the website, invite and add members to the organization, and manage resources.

- Learn essential tasks and best practices for setting up ArcGIS Online with the *ArcGIS Online Implementation Guide*.

ArcGIS Pro

ArcGIS Pro can help you explore, visualize, and analyze data; create 2D maps and 3D scenes; and share your work to your ArcGIS Online or ArcGIS Enterprise portal.

- Learn essential tasks for getting your organization started with ArcGIS Pro with the *ArcGIS Pro Implementation Guide*.

Identify foundational data

Gather and map foundational data for areas covered by your organization. These layers include basic infrastructure and administrative areas:

- Administrative and jurisdictional boundaries, such as city and country boundaries
- Population and demographics
- Infrastructure (roads, bridges, dams, utilities, communications, and so on)
- Major facilities and landmarks
- Water features (lakes, streams, rivers, and so on)
- Remote sensing data such as airborne and spaceborne imagery and other variable data
- Earth sciences domain-specific data

Add ready-to-use content from ArcGIS Living Atlas of the World, which contains several live feeds that provide real-time information that can be used in addition to local data:

- Weather feeds
- Disaster feeds
- Earth observation feeds
- Multispectral feeds

Discover more live feeds in ArcGIS Living Atlas at links.esri.com/atlas_live.

Analysis

R Project for Statistical Computing

R, also known as the R Project for Statistical Computing, is the prevalent and fastest-growing environment for statistical computing, including environmental sciences. As a member of the R Consortium, Esri is part of the R community and supports the R project. The new R–ArcGIS Bridge, now compatible with Microsoft R Open, is advancing the capabilities of analyses across disciplines. R–ArcGIS Bridge lets you combine statistical models with R and ArcGIS for spatial data access, visualization, and analysis. This combination can extend R models by using ArcGIS to access data, spatial algorithms, and more. ArcGIS software offers performance, scalability, and spatial validation when users build R-based analytics tools in an R or ArcGIS environment.

Jupyter integration

R–ArcGIS Bridge in Jupyter notebooks allows you to document your analysis by combining code with visuals and text.

ArcGIS Notebooks: Spatial analysis meets data science

ArcGIS Notebooks offers a Jupyter notebook experience for spatial analysis. Notebooks offers an array of capabilities:

- **Enhance your data engineering:** Prepare your data using spatial and open-source libraries to isolate an area of interest and identify anomalies. Add context with location data.
- **Optimize your analyses with location:** Go beyond proximity with advanced analytics. Arrive at statistically significant results with spatial algorithms and open-source Python libraries.

- **Predict with spatial AI:** Train and perform inference on models using built-in tools with machine learning and deep learning frameworks to solve complex spatial problems.
- **Find efficiencies and reduce errors:** Use the built-in ArcGIS Python Libraries to automate and schedule repetitive administrative tasks in your Web GIS.
- **Reproducible research across teams:** Explain your analyses by combining Python with interactive visuals and descriptions. Share a workspace that is familiar to other data scientists.

Exploring open-source tools using ArcGIS

ArcGIS enables users to integrate more easily within their specific IT contexts to use the power of location and data from any source. These tools support users ranging from large enterprise systems integrators and developers seeking to build apps to someone who wants to share information and collaborate with others. Open-source tools are also available on GitHub.

Data management

You can prepare data using spatial and open-source libraries to isolate an area of interest, identify anomalies, and add context with location data:

- **Add data:** Access your data and contextualize it with datasets built by the ArcGIS community and Esri's team of cartographers, demographers, and statisticians.

- **Visualize:** Dynamically create maps and visualizations that tell a story. Explore data distributions, find patterns, and view results in 2D and 3D.
- **Analyze:** Combine spatial analytics from ArcGIS with open-source Python libraries to solve complex problems and build precise models.

Collaboration

ArcGIS software helps you adopt a more inclusive approach to science and innovation with a framework that engages departments, science disciplines, and the public:

- Make your models accessible and easy to find, increasing transparency and efficiency.
- Work together with other analysts and data scientists.
- Integrate analytics across your organization.
- Infuse analytics into your organization's decision-making by making your models enterprise-ready and accessible.
- Deploy your analysis as ready-to-run tools or as notebooks from your workstation, in the cloud, or behind your security system.
- Run your analytic models on a schedule so your organization operates on the most up-to-date data.
- Turn sophisticated analysis into compelling stories that incorporate text, images, and videos.
- Share your narratives with colleagues, clients, and the public.

Learning by doing

Direct learning will strengthen your understanding of GIS and how it can be used. Esri offers a collection of free online story-driven lessons that allow you to experience GIS when it is applied to real-life problems.

Learning the basics

To start your GIS journey, you can review these examples:

- **Create a map:** Create a web map with ArcGIS Online.
- **Create an app:** Configure and share an app that puts your web map to greater use.
- **Get started with ArcGIS Online:** Get an introduction to web mapping with ArcGIS Online.
- **Get started with ArcGIS Pro:** Learn the basics of ArcGIS Pro.

Earth sciences–specific tutorials

ArcGIS tutorials directly applicable to earth sciences include these examples:

- **Perform a site suitability analysis for a new wind farm:** Determine the optimal location for a set of new high-efficiency wind turbines in Colorado.
- **Monitor forest change over time:** Detect and analyze forest disturbance and recovery from a Landsat time series in the West Cascades ecoregion in Oregon.

- **Get started with multidimensional multispectral imagery:** Use a multidimensional stack of Landsat imagery to visualize how a Chilean copper mine has changed over time.
- **Explore and animate geological data with voxels:** View and analyze multidimensional soil voxels for the Netherlands.
- **Assess hail damage in cornfields with satellite imagery in ArcGIS Pro:** Compute the change in vegetation before and after a hailstorm.
- **Identify groundwater vulnerable areas:** Locate and map areas that are prone to groundwater contamination.
- **Map the effects of climate change on the ocean:** Map warming temperatures, acidification, and deoxygenation with climate data.

Learn more

For links to examples shown in this book and additional resources, visit the book web page:

go.esri.com/gfes-resources

CONTRIBUTORS

Matt Ball
Jim Baumann
Chris Chiappinelli
Monica Pratt
Jill Saligoe-Simmel
Benjamin J. Smith
Citabria Stevens
Carla Wheeler

ABOUT ESRI PRESS

ESRI PRESS IS AN AMERICAN BOOK PUBLISHER AND PART OF Esri, the global leader in geographic information system (GIS) software, location intelligence, and mapping. Since 1969, Esri has supported customers with geographic science and geospatial analytics, what we call The Science of Where®. We take a geographic approach to problem-solving, brought to life by modern GIS technology, and are committed to using science and technology to build a sustainable world.

At Esri Press, our mission is to inform, inspire, and teach professionals, students, educators, and the public about GIS by developing print and digital publications. Our goal is to increase the adoption of ArcGIS and to support the vision and brand of Esri. We strive to be the leader in publishing great GIS books, and we are dedicated to improving the work and lives of our global community of users, authors, and colleagues.

Related titles

GIS for Science, Volume 3: Maps for Saving the Planet

Dawn J. Wright & Christian Harder (eds.)

9781589486713

Top 20 Essential Skills for ArcGIS Pro

Bonnie Shrewsbury and Barry Waite

9781589487505

Mapping America's National Parks: Preserving Our Natural and Cultural Treasures

U.S. National Park Service

9781589485464

Local Voices, Local Choices: The Tacare Approach to Community-Led Conservation

Jane Goodall Institute

9781589486461